LÁSZLÓ LOVÁSZ
Eötvös Loránd University, Budapest, Hungary

An Algorithmic Theory of Numbers, Graphs and Convexity

SOCIETY FOR INDUSTRIAL A
PHILADELPHIA, PENNSYLVANIA

Second printing 1989.
Library of Congress Catalog Card Number 86-61532.

ISBN 0-89871-203-3.

Printed for the Society for Industrial and Applied Mathematics by Capital City Press, Montpelier, Vermont.

Contents

1 Introduction

Chapter 1. How to Round Numbers

7 1.0. Preliminaries: On algorithms involving numbers.
9 1.1. Diophantine approximation. Problems.
15 1.2. Lattices, bases, and the reduction problem.
28 1.3. Diophantine approximation and rounding.
34 1.4. What is a real number?

Chapter 2. How to Round a Convex Body

41 2.0. Preliminaries: Inputting a set.
42 2.1. Algorithmic problems on convex sets.
47 2.2. The Ellipsoid Method.
54 2.3. Rational polyhedra.
58 2.4. Some other algorithmic problems on convex sets.
61 2.5. Integer programming in a fixed dimension.

Chapter 3. Some Applications in Combinatorics

65 3.1. Cuts and joins.
75 3.2. Chromatic number, cliques, and perfect graphs.
82 3.3. Minimizing a submodular function.

87 References

Introduction

There is little doubt that the present explosion of interest in the algorithmic aspects of mathematics is due to the development of computers — even though special algorithms and their study can be traced back all the way through the history of mathematics. Mathematics started out in Egypt and Babylon as a clearly algorithmic science. In ancient Greece the foundations of its "descriptive" or "structural" line were established; but even here we find algorithmic problems — just think of the problem of constructibility of various geometric figures by compass and ruler. I find it amazing that this problem was precisely formulated in the framework of axiomatic geometry (reflecting the current state of devices at the disposal of the Greeks when they were carrying out those constructions). It is unnecessary to say how much this problem contributed to the later development of geometry and to the creation of algebra: both the positive and the negative results inspired fundamental notions and theorems (e.g. the golden ratio on the one hand and the solvability of algebraic equations by radicals, in particular by square roots, on the other).

In our day, the development of computers and the theory of algorithms and their complexity have produced a similar situation. In the last centuries, a vast body of "structural" mathematics has evolved. Now that we are interested in the algorithmic aspects of these results, we meet extremely difficult problems. Some of the most elementary results in number theory, geometry, algebra, or calculus become utterly difficult when we ask for algorithms to find those objects whose existence is (at least by now) easily established. Just think of the elementary fact, known to Euclid, that any integer has a unique prime factorization, and contrast it with the apparent intractability of the corresponding algorithmic problem, namely, the problem of *finding* this decomposition.

One may remark at this point that there is a trivial way to factor a natural number: just test all natural numbers up to its square root to find a proper divisor (if any exists). To formulate the factorization problem precisely, one has to explain why this solution is unsatisfactory. More generally speaking, one has to introduce the right models of computation, observing the resources (time, space), and contrast them with examples which are considered "good" or "bad" on a practical basis. This approach has shown that one measure of efficiency, namely polynomial time, accords particularly well with subjective feelings and with rules–of–thumb concerning which problems are inherently "easy" or "difficult" and which algorithms are "superficial" or "deep". This fact justifies a thorough study of polynomial time algorithms. (This is so in spite of the – just – criticism which has been directed against the identification of polynomiality with real world computational efficiency or practicality.) The frames of these notes do not allow one to give a precise introduction to the theory of computational complexity. We accept polynomial time (i.e. running time bounded by a polynomial of the length of the input) as our main criterion for the "goodness" of an algorithm (from a theoretical point of view), and study the class P of problems solvable in polynomial time. A (probably) wider class of problems is the class NP , solvable in polynomial time by a non–deterministic Turing machine. The rather technical exact definition of this notion is not essential for the understanding of this book; let it suffice to say at this point that almost all combinatorial decision problems (e.g. planarity or 4–colorability of a graph, the existence of a perfect matching or Hamilton circuit, etc.) belong to this class. The problem of whether $P = NP$ is one of the most outstanding unsolved problems in mathematics nowadays.

The class NP contains some problems, called NP*–complete*, to which every other problem in NP can be reduced. So if $P \neq NP$ then no NP–complete problem can be solved in polynomial time. If a problem is not in NP (say, if it is not a decision problem), but is at least as hard as some NP–complete problem, then it is called NP–hard. We shall assert that certain problems are NP–complete or NP–hard; the practical implication of this is that most probably there is no polynomial time algorithm to solve such a problem at all.

With this framework of computational complexity theory at hand, a very wide scientific program arises: to study, or rather to re-study, the results of classical "structural" mathematics from the point of view of the polynomial time computability of the results. Very little of this program has been carried out. We have a fairly good understanding of the algorithmic aspects of fundamental problems in graph theory and combinatorial optimization, and a deep and promising entry into some parts of number theory, group theory and geometry; we know very little about the complexity aspects of algorithmic algebra in general or about algorithmic analysis.

I would not find it fortunate to consider such a program as a thesis about the uselessness of "structural" mathematics. Rather, it ought to fertilize classical mathematics with new problems and application possibilities. On the one hand, the design of better and better algorithms requires more and more knowledge of classical mathematics. On the other, the proof of negative results (e.g. $P \neq NP$) will certainly require the development of entirely new structural results. (Think of the proof of the non–constructibility of, say, the regular heptagon: this required the development of analytic geometry and algebra!)

It is of course possible that the farther away we get from the sources of the notion of polynomial time, namely from combinatorial algorithms, the less significant this notion becomes. But it is also true that some algorithms, which were developed just from a theoretical point of view to resolve the question of polynomial–time tractability of some rather special problems, have turned out to have impact on much wider and more fundamental questions; some of those algorithms appear to have practical applications as well. In these lecture notes we study the implications of two such algorithms: the Ellipsoid Method and the simultaneous diophantine approximation algorithm.

The Ellipsoid Method arose as a method in non–linear optimization, but became widely known after it was applied by Khachiyan (1979) to solve an important, though mainly theoretical, problem, namely the polynomial time solvability of linear programs. (This is the problem of finding a solution to a system of linear inequalities in several variables that maximizes a given linear function.) Since then, it has been shown (Grötschel, Lovász and Schrijver (1981,1984b)) that it provides a very general algorithmic principle, which is basic in algorithmic geometry and polyhedral combinatorics.

From a practical point of view the Ellipsoid Method has proved a failure, but, recently, Karmarkar (1984) developed another method to solve linear programs in polynomial time which appears to be even practically competitive with the Simplex Method. Karmarkar's method is based on different ideas, but some influence of the Ellipsoid Method on his approach cannot be denied.

It is an important, though simple, fact from geometry that a closed convex set can be defined as the convex hull of its extremal points, but also as the intersection of its supporting halfspaces. One of the main consequences of the Ellipsoid Method is that these two, in a sense dual, representations of a convex set are not only logically but also algorithmically equivalent. This fact from algorithmic geometry, in turn, has numerous consequences in combinatorics.

These combinatorial applications depend on another line of development in the foundations of combinatorics, namely, polyhedral combinatorics. A combinatorial optimization problem (e.g. finding a shortest path, a maximum matching, a shortest traveling salesman tour, etc.) has typically only finitely many possible solutions, so from a "classical" mathematical point of view, there is no problem here: one may inspect these solutions (all paths, all matchings, etc.) one by one, and select the best. The problem becomes non–trivial only when we try to achieve this with much less work. This is only possible, of course, because the set of solutions and values of the objective function is highly structured. But how to formalize this "structuredness" in a way which makes the problem tractable? One way to do so is to consider the convex hull of incidence vectors of solutions. If the polyhedron obtained this way has a "nice" description as the solution set of a system of linear inequalities, then one may use the powerful results and methods from linear programming. So, for example, we may use the Duality Theorem to obtain a "good characterization" of the optimum value. There is, however, an important difficulty with this approach if we wish to use it to obtain algorithms: the number of linear inequalities obtained this way will in general be exponentially large (in terms of the natural "size" of the problem). Nevertheless, good polyhedral descriptions and efficient algorithmic solvability have usually gone hand in hand. Sometimes, a good polyhedral description follows from efficient algorithms. In other cases, the methods developed to obtain polyhedral results (or, more or less equivalently, minimax results and good characterizations of various properties) could be refined to obtain efficient algorithms. Polyhedral descriptions are also used (and with increasing success) in obtaining practical algorithms for otherwise intractable combinatorial optimization problems. But it was the Ellipsoid Method which provided the first exact results in this direction (Karp and Papadimitriou (1980), Grötschel, Lovász and Schrijver (1981), Padberg and Rao (1982)). The algorithmic equivalence of the two descriptions of convex sets mentioned above implies the following. Suppose that we find an (implicit) description of a polyhedron by linear inequalities, which is "nice" enough in the sense that there exists a polynomial–time procedure to check if a given vector satisfies all these inequalities. Then we can optimize any linear objective function over this polyhedron in polynomial time.

It turns out that a very large number of combinatorial optimization problems can be shown to be polynomially solvable using this principle. There are at least two very important problems (finding a maximum independent set in a perfect graph, and minimizing a submodular function) whose polynomial time solvability is known only via this principle.

Note that I have tried to word these implications carefully. The polynomial time algorithms which follow from these general considerations are very far from being practical. In fact, often the Ellipsoid Method, which is quite impractical in itself, has to be called as a subroutine of itself, sometimes even nested three or four times. This ridiculous slowness is of course quite natural for a method which can be applied in such a generality – and hence makes so little use of the specialities of particular problems. Once polynomially solvable and NP–hard problems have been "mapped out", one can try to find for those known to be polynomially solvable special–purpose algorithms which solve them really efficiently. Such research has indeed been inspired by "ellipsoidal" results, e.g. in connection with the problem of minimizing submodular functions.

Polyhedral combinatorics also shows that combinatorial optimization problems are special cases of the integer linear programming problem (a variation of the linear programming problem, where the variables are constrained to integers). There is, however, no hope of obtaining a polynomial–time algorithm for the integer linear programming problem in general, since it is known to be NP–complete. On the other hand, a celebrated result of H.W. Lenstra, Jr. (1983) shows that for every fixed dimension, the integer linear programming problem can be solved in polynomial time (of course, the polynomial in the time bound depends on the dimension).

A most significant aspect of Lenstra's result is that it connects ideas from number theory and optimization. If one formulates them generally, integer linear programming and the geometry of numbers both study the existence of integral points in (usually convex) bodies, but traditional conditions, concrete problems and solution methods in both fields are very different. Lenstra's work applied ideas from the theory of diophantine approximation to integer linear programs. This approach, as we shall see, has led to many further results in optimization.

The second main ingredient of these notes is an algorithm for finding good approximations to a set of real numbers by rationals with a common denominator. This classical problem of diophantine approximation has come up – somewhat unexpectedly – in settling a minor problem left open by the first results on the Ellipsoid Method, namely extending the results to the case of non–full–dimensional polyhedra. Using ideas of H.W. Lenstra's result mentioned above, A.K. Lenstra, H.W. Lenstra and L. Lovász (1982) obtained a polynomial–time algorithm for simultaneous diophantine approximation. Since then, this algorithm has been applied to rather different problems in algebra, number theory, and combinatorics. It has turned out that it nicely complements the Ellipsoid Method in more than one case.

The organization of these notes is as follows. We begin with a discussion of a very fundamental problem in numerical mathematics, namely rounding. It turns out that the obvious way of rounding the data of a problem, namely one by one, has some deficiencies from a theoretical point of view. More sophisticated (but still polynomial–time) rounding procedures can be worked out, which satisfy rather strong requirements. These procedures depend on simultaneous diophantine approximation, and we shall discuss the algorithm of Lenstra, Lenstra and Lovász in detail. We also go into several applications, refinements and analogs of this algorithm. These considerations also lead to a certain model of computations with real numbers. Roughly speaking, we identify every real number with an "oracle", which gives rational approximations of the number in question with arbitrary precision. It is surprising and non–trivial that (under appropriate side–constraints), for rational and algebraic numbers, this "oracle" model turns out to be polynomial–time equivalent to the natural direct encodings.

Then we turn to the discussion and applications of the Ellipsoid Method. We develop an "oracle" model for an algorithmic approach to convex bodies, analogous to, and in a sense generalizing, the "oracle" model of real numbers. We prove that a number of quite different descriptions of a convex body are polynomial–time equivalent. The methods also yield a way to compute reasonable upper and lower bounds on various measures of convex bodies, such as their volumes, widths, diameters etc. The oracle model we use enables us to prove that these measures cannot be exactly determined in polynomial time. This approach was worked out by Grötschel, Lovász and Schrijver (1981, 1984a).

The two main algorithms come together when we study polyhedra with rational vertices. If we know an upper bound on the complexity of the vertices (as is virtually always the case e.g. for combinatorial applications), then various outputs of the Ellipsoid Method can be made "nicer" by rounding. So, for example, "almost optimizing" points and "almost separating" hyperplanes can be turned into optimizing vertices and separating facets. Similar methods were very recently applied by A. Frank and É. Tardos (1985) to show that many (if not most) combinatorial optimization algorithms can be turned from polynomial into strongly polynomial (i.e. the number of arithmetic operations they perform can be made independent of the size of the input numbers). A nice special case is the result found by Tardos that the minimum cost flow problem can be solved in strongly polynomial time, which settled a long–standing open problem. Furthermore, we show how a similar combination of previous ideas leads to a sharpened version of Lenstra's result on integer programming in bounded dimension: it turns out that the hard part of the work can be done in time which is polynomially bounded even if the dimension varies, and the problem can be reduced to distinguishing a finite number of cases, where this number depends only on the dimension (but, unfortunately, exponentially).

We conclude these notes by surveying some applications of these results to combinatorial optimization. We do not attempt to be comprehensive, but rather to study three topics: first, cut problems (including flows, matchings, Chinese Postman tours etc.); second, optimization in perfect graphs; and third, minimization of submodular functions. The moral of the first study is that using the equivalence principles supplied by the Ellipsoid Method, linear programming duality, the Greedy algorithm and some trivial reductions among combinatorial optimization problems, a large number of very different combinatorial optimization problems can be shown polynomial–time solvable. The second and the third applications are interesting because so far the only way to show their polynomial–time solvability is by the Ellipsoid Method.

Throughout these notes, I have put emphasis on ideas and illustrating examples rather than on technical details or comprehensive surveys. A forthcoming book of M. Grötschel, L. Lovász and A. Schrijver (1985) will contain an in–depth study of the Ellipsoid Method and the simultaneous diophantine approximation problem, their versions and modifications, as well as a comprehensive tour d'horizon over their applications in combinatorics.

Thus a large part of these notes is based on that book and on many discussions we had while writing it, and I am most indebted to my co–authors for these discussions and for their permission to include much of our common work in these notes. Another basis of my work was the series of lectures I gave on this topic at the AMS–CBMS regional conference in Eugene, Oregon, August 1984. I have made use of many remarks by my audience there.

I am also most grateful to the Institute of Operations Research of the University of Bonn for both providing ideal circumstances for my work on this topic and for the technical assistance in the preparation of these notes. My thanks are due to Frau B. Schaschek for the careful typing into TEX.

CHAPTER 1

How to Round Numbers

1.0. Preliminaries: On algorithms involving numbers.

The most common objects of mathematics (be it algorithmic or descriptive) are numbers. It is in fact somewhat difficult to convince a non-mathematician that mathematics is concerned with many other kinds of objects: groups, geometric configurations, graphs, etc. It may be partly because everybody feels very familiar with numbers that the role of numbers in various algorithms is often not very clearly defined, and that there are various models of computation with numbers.

Computers typically have a fixed number of bits reserved for each number, and this precision is sufficient for most real-world problems. Therefore, the number of digits rarely concerns a practicing programmer. However, there are some reasons why computing with many digits is becoming important:

> Multiprecision computation (if it can be done cheaply) may be a cure to the instability of instable but otherwise favorable algorithms.
>
> In the solution of a problem in which all inputs and outputs are small numbers, information generated during the computation may be stored in the form of long numbers. Later on we shall discuss how the computation of a root of a polynomial with high precision can be a step in finding its factorization into irreducible polynomials.
>
> Number theoretic algorithms (primality testing, factoring an integer, etc.) are interesting for integers with 100 to several hundred digits. These studies, in turn, have important applications in cryptography, random number generation and communication protocols. Alas, we cannot go into this lovely area here.

To get a hardware-independent model of numbers in algorithms, two natural possibilities arise:

I. We can take into account the number of digits in, say, the binary expansion of the numbers. This means that if an integer n occurs in the input, it adds

$$\langle n \rangle = 1 + \lceil \log_2(n+1) \rceil$$

to the input size (this is the number of binary digits of n, plus 1 for the sign if n is non-zero). Arithmetic operations like addition, multiplication and comparison of two numbers must be carried out bitwise; so the multiplication of two k-digit numbers by the method learned at school adds about k^2 to the running time. In this model, rational numbers are encoded in the form $r = p/q$, where g.c.d. $(p, q) = 1$ and $q > 0$. It is convenient to extend the "input size" notation to rational numbers by letting $\langle r \rangle = \langle p \rangle + \langle q \rangle$. If $a = (a_1, \dots, a_n) \in \mathbb{Q}^n$ is a vector, we set

$$\langle a \rangle = \langle a_1 \rangle + \ldots + \langle a_n \rangle .$$

The input size $\langle A \rangle$ of a rational matrix A is defined analogously. It is clear that for any integer n, $|n| < 2^{\langle n \rangle}$. It is easy to see that if $r \in \mathbb{Q}$, $r > 0$ then

$$2^{-\langle r \rangle} < r < 2^{\langle r \rangle} .$$

It is also easy to extend these estimates to vectors, matrices etc. An inequality we shall need is the following: if A is a non-singular $n \times n$ matrix then

$$2^{-\langle A \rangle} < \det A < 2^{\langle A \rangle} .$$

We shall refer to this model as the *binary encoding*. If we do not say specifically otherwise, we always mean this model.

Sometimes $|n|$ is also used as a measure of how much the integer n contributes to the input size. It is said then that n is in *unary encoding*. Measuring the input size in unary encoding makes the algorithms seemingly faster. An algorithm which is polynomial if the input data are given in the unary encoding, but not if they are given in the binary encoding, is often called *pseudopolynomial*. This notion, however, shall not concern us; we only use "unary encoding" as a convention to measure the contribution of an input number to the size of the input.

II. An important alternative, however, is to say that one integer contributes one to the input. Addition, multiplication and comparison of two integers count then as one step. When we say that the determinant of an $n \times n$ matrix can be determined in $O(n^3)$ time, we mean it in this sense; if bitwise operations were counted then, of course, the running time would depend on the lengths of the entries of the matrix. We shall call this model of computation the *arithmetic encoding*.

An algorithm may be polynomial in the binary sense but not in the arithmetic sense; e.g. the computation of the g.c.d. of two numbers, or the binary search. It may also happen conversely: n times repeated squaring takes $O(n)$ time in the arithmetic sense, but only to write down the result takes exponential time in the binary sense. But this is essentially the only thing that can go wrong with an algorithm that uses only a polynomial number of arithmetic operations. If we can prove that, besides, all numbers occurring during the run of the algorithm have input size bounded by polynomial of the original input size, then the algorithm will be also polynomial in the binary sense. We shall call an algorithm *strongly polynomial* if it takes polynomial time in the arithmetic sense, and only takes polynomial space in the binary encoding. Every strongly polynomial algorithm is polynomial.

So far, we have only discussed computations with integers (and with rational numbers, which are encoded as pairs of integers). It is questionable whether one should bother with developing any model of real (or complex) number computations; obviously, all inputs and outputs of algorithms are necessarily rational. But there are problems whose solution, at least in the theoretical sense, is an irrational number (see e.g. Section 3.2). There are also some counterexamples that involve irrational numbers, e.g. Ford and Fulkerson's example for the non–termination of the flow algorithm. One feels that such examples do say something about the real–word behaviour of these algorithms – but what?

In Section 1.4 we sketch a model for real number computations, based on an "oracle"–approach. The main goal of this is not to propose new foundations for "constructive analysis", but rather to point out two connections with the main topic of this book. First, the role of rational and algebraic numbers in this model can be very nicely described using the simultaneous diophantine approximation techniques developed in this chapter. Second, we shall extend this "oracle"–approach to higher dimensions in Chapter 2, and this will provide the framework for our algorithmic treatment of convex sets. Another oracle model will be used in Chapter 3 to describe algorithms for matroids and submodular functions (in this context, oracles have been used for a long time).

1.1. Diophantine approximation. Problems.

In a general sense, this chapter is about how to "round" numbers. To "round" a number $\alpha \in \mathbb{R}$ means that we replace it by a rational number which is of a sufficiently simple form and at the same time sufficiently close to α . If we prescribe the denominator q of this rational number p/q , then the best choice for p is either $p = \lfloor \alpha q \rfloor$ or $p = \lceil \alpha q \rceil$. This is the most common way of rounding a number; usually $q = 10^k$ for some k . The error resulting from such a rounding is

$$\left| \alpha - \frac{p}{q} \right| \le \frac{1}{2q} . \tag{1.1.1}$$

We shall find, however, that often this approximation is not good enough. A classical result of Dirichlet says that if we do not prescribe the denominator, but only an upper bound Q for it, then there always exists a rational number p/q

such that

$$\left|\alpha - \frac{p}{q}\right| \le \frac{1}{Qq}, \quad 0 < q \le Q \,. \tag{1.1.2}$$

There also exists a classical method to find such a rational number p/q : this is the so-called *continued fraction expansion* of α . One can write α uniquely in the form

$$\alpha = a_0 + \cfrac{1}{a_1 + \cfrac{1}{a_2 + \dots}}$$

where $a_0 = \lfloor \alpha \rfloor$ and $a_1, a_2, \dots$ are positive integers. In fact, $a_1, a_2, \dots$ can be defined by the recurrence

$$\alpha_0 = \alpha, \quad a_0 = \lfloor \alpha \rfloor, \quad \alpha_{k+1} = \frac{1}{\alpha_k - a_k}, \quad a_{k+1} = \lfloor \alpha_{k+1} \rfloor \,.$$

For an irrational number α , this expansion is infinite; for a rational number α , it is finite but may be arbitrarily long. If we stop after the kth a_k , i.e. determine the rational number

$$\frac{g_k}{h_k} = a_0 + \cfrac{1}{a_1 + \cfrac{1}{\ddots \; a_{k-1} + \cfrac{1}{a_k}}}$$

then we obtain the kth *convergent* of α . It is known that

$$\frac{g_k}{h_k} \to \alpha \quad (k \to \infty) \,;$$

in fact,

$$\left|\alpha - \frac{g_k}{h_k}\right| \le \frac{1}{h_k h_{k+1}} \tag{1.1.3}$$

and h_k grows exponentially fast, so the right hand side of (1.1.3) tends to 0. If we let k be the largest subscript for which $h_k \le Q$, then $h_{k+1} > Q$ and so

$$\left|\alpha - \frac{g_k}{h_k}\right| < \frac{1}{Qh_k} \,,$$

i.e. we have found an approximation satisfying (1.1.2).

The continued fraction expansion has many nice properties. The numbers g_k and h_k satisfy the recurrence

$$g_{k+1} = a_{k+1}g_k + g_{k-1}, \qquad h_{k+1} = a_{k+1}h_k + h_{k-1} \ ,$$

which makes them easily computable. They satisfy the important identity

$$g_{k+1}h_k - g_k h_{k+1} = (-1)^k \ .$$

One can use continued fractions to find approximations of various kinds. For example, consider the following problem:

(1.1.4) Best approximation problem.
Given a number $\alpha \in \mathbb{Q}$, and an integer $Q > 0$, find a rational number p/q such that $0 < q \leq Q$ and $|\alpha - \frac{p}{q}|$ is as small as possible.

We can solve (1.1.4) as follows (see e.g. Khintchine (1935)). Let, as before, g_k/h_k be the last convergent of α with $h_k \leq Q$. Then g_k/h_k is a very good approximation of α , but it may not be the best. It is certainly the best if $\alpha = g_k/h_k$, so we may assume that this is not the case. To find the best approximation, compute the largest $j \geq 0$ such that

$$h_{k+1} + jh_k \leq Q \ .$$

Then either $\frac{g_{k-1}+jg_k}{h_{k-1}+jh_k}$ or $\frac{g_k}{h_k}$ is the rational number with denominator at most Q closest to α . Let us assume e.g. that

$$\frac{g_{k-1}}{h_{k-1}} < \alpha < \frac{g_k}{h_k}$$

(i.e. k is odd). Then we also have that

$$\frac{g_{k+1}}{h_{k+1}} = \frac{g_{k-1} + a_{k+1}g_k}{h_{k-1} + a_{k+1}h_k} < \alpha \ .$$

By the choice of k , we have that $h_{k+1} = h_{k-1} + a_{k+1}h_k > Q$ and hence by the choice of j , we have $0 \leq j < a_{k+1}$. Hence $\frac{g_{k-1}+jg_k}{h_{k-1}+jh_k}$ lies between $\frac{g_{k-1}}{h_{k-1}}$ and $\frac{g_{k+1}}{h_{k+1}}$ and so in particular $\frac{g_{k-1}+jg_k}{h_{k-1}+jh_k} < \alpha$.

Now if we show that every rational number in the interval $(\frac{g_{k-1}+jg_k}{h_{k-1}+jh_k}, \frac{g_k}{h_k})$ has denominator $> Q$, then it follows that one of the endpoints of this interval must be the rational number closest to α with denominator $\leq Q$.

So let

$$\frac{g_{k-1} + jg_k}{h_{k-1} + jh_k} < \frac{r}{s} < \frac{g_k}{h_k},$$

where $r, s \in \mathbb{Z},\ s > 0$. Then clearly

$$\frac{r}{s} - \frac{g_{k-1} + jg_k}{h_{k-1} + jh_k} \geq \frac{1}{s(h_{k-1} + jh_k)}$$

and

$$\frac{g_k}{h_k} - \frac{r}{s} \geq \frac{1}{sh_k} \ .$$

On the other hand,

$$\frac{g_k}{h_k} - \frac{g_{k-1} + jg_k}{h_{k-1} + jh_k} = \frac{g_k h_{k-1} - h_k g_{k-1}}{h_k(h_{k-1} + jh_k)} = \frac{1}{h_k(h_{k-1} + jh_k)} \ .$$

So

$$\frac{1}{s(h_{k-1} + jh_k)} + \frac{1}{sh_k} \leq \frac{1}{h_k(h_{k-1} + jh_k)}$$

and hence

$$s \geq h_{k-1} + (j+1)h_k > Q \ .$$

This shows that r/s is no solution to our problem.

Sometimes this problem arises in a "reverse" form: (*) Given $\alpha \in \mathbb{Q}$ and $\epsilon \in \mathbb{Q},\ \epsilon > 0$, find a rational number $\frac{p}{q}$ such that $q > 0, \cdot |\alpha - \frac{p}{q}| < \epsilon$ and q is as small as possible. These two problems are however in a sense equivalent. This is best seen by noticing that both are equivalent to the following: (**) Given $\alpha \in \mathbb{Q},\ \epsilon \in \mathbb{Q},\ \epsilon > 0$ and $Q \in \mathbb{Z},\ Q > 0,$ decide if there exist integers p, q such that $0 < q \leq Q$ and $|\alpha - \frac{p}{q}| < \epsilon$ and find such integers if they exist. If we can solve the Best Approximation Problem, then of course we can also solve (**) by simply checking to see if the rational number p/q with $0 < q \leq Q$ and closest to α satisfies $|\alpha - \frac{p}{q}| < \epsilon$. Conversely, if we can solve (**) then by binary search over the interval $(0,1)$, we can find the smallest ϵ for which (**) has a solution and so we can also find the best approximation of α with denominator $\leq Q$. A similar argument can be used to show the equivalence of (*) and (**).

Suppose now that we have several numbers $\alpha_1, \ldots, \alpha_n$ to round. We may do this individually; however, some unpleasant things may happen. For example, if we want to round $\alpha_1 = 1.5$ and $\alpha_2 = 1.5$ to the next integer, then we may obtain $\alpha_1 \sim 1$ and $\alpha_2 \sim 2 \neq 1$. So equality may not be preserved by rounding! Of course, we may add some artificial conventions to break the ties (e.g. round "up" in such cases). But then consider $\alpha_1 = 0.1422,\ \alpha_2 = 0.2213$ and $\alpha_3 = 0.6359$ (so that $\alpha_1 + \alpha_2 + \alpha_3 = 1$). Choose some Q , say $Q = 100$, and compute the best approximations of α_1, α_2 and α_3 by rational numbers with denominators ≤ 100 . One obtains

$$\alpha_1 \sim \frac{1}{7} = 0.1428\ldots, \qquad \alpha_2 \sim \frac{2}{9} = 0.2222\ldots, \qquad \alpha_3 \sim \frac{7}{11} = 0.6363\ldots \ .$$

These numbers are indeed much nicer, but we did lose the property that $\alpha_1 + \alpha_2 + \alpha_3 = 1$; in fact, $\frac{1}{7} + \frac{2}{9} + \frac{7}{11} \neq 1$. In a sense, the main purpose of this chapter is to give a simultaneous rounding procedure in which such a simple relation among the numbers is never lost. We shall, however, postpone the exact formulation of this property to Section 1.4.

The main trick is to use a rounding procedure which results in rational approximations $\frac{p_1}{q}, \ldots, \frac{p_n}{q}$ with the same denominator. If we prescribe the denominator, the trivial error (1.1.1) can of course be achieved. But we need a rounding procedure that gives better approximation, its expense being that only an upper bound on the (common) denominator can be prescribed.

(1.1.5) Simultaneous diophantine approximation problem.
Given $\alpha_1, \ldots, \alpha_n \in \mathbb{Q}$, $\epsilon > 0$, and $Q > 0$, find integers $p_1, \ldots, p_n$ and q provided

$$0 < q \leq Q, \quad \left|\alpha_i - \frac{p_i}{q}\right| \leq \frac{\epsilon}{q} \quad (i = 1, \ldots, n) .$$

If $\epsilon = \frac{1}{2}$, then this problem is trivially solvable, even if we prescribe q ; for $\epsilon < \frac{1}{2}$, however, we have to choose an appropriate q . How large must this q be? Dirichlet's theorem extends to this general case and asserts that there always exists a solution of (1.1.5) provided

$$Q \geq \epsilon^{-n} . \tag{1.1.6}$$

However, contrary to the case of a single α , no efficient algorithm is known for finding such an approximation. The main result of this chapter will be a polynomial–time algorithm that solves (1.1.5) provided

$$Q \geq 2^{n^2} \epsilon^{-n} . \tag{1.1.7}$$

The bound of 2^{n^2} is rather large. However, in many applications the bound on q does not play any role, as long as q can be bounded independently of $\alpha_1, \ldots, \alpha_n$ and the number of its digits is polynomially bounded.

A closely related problem is the following (in fact it is easy to find a common form of the two problems (see the next section)):

(1.1.8) Small integer combination problem.
Given $\alpha_0, \alpha_1, \ldots, \alpha_n \in \mathbb{Q}$, and $\epsilon, Q > 0$, find integers $q_0, q_1, \ldots, q_n$, not all 0, such that

$$\left|\sum_{i=0}^{n} q_i \alpha_i\right| \leq \epsilon$$

and $q_i \leq Q \quad (i = 1, \ldots, n)$.

It is customary to assume that $\alpha_0 = 1$; this is, of course, no restriction of generality. Also note that no upper bound on q_0 is prescribed; indeed a bound on q_0 follows trivially from the others. (The more general problems discussed in the next section will also contain the version when bounds are given for all the q_i's.)

An analogue of Dirichlet's theorem tells us that there always exists a solution of this problem provided $Q \geq \epsilon^{-1/n}$. Again, no efficient algorithm is known to solve this problem; however, we shall be able to find a solution provided $Q \geq 2^n \epsilon^{-1/n}$, which will be good enough for our applications.

Let us briefly mention the problem of *inhomogeneous diophantine approximation.* In this case, we are given $2n$ numbers $\alpha_1, \ldots, \alpha_n, \beta_1, \ldots, \beta_n$, and $\epsilon, Q > 0$, and we are looking for integers $p_1, \ldots, p_n$, q such that

$$|q\ \alpha_i - p_i - \beta_i| \leq \epsilon$$

and $0 < q \leq Q$.

Contrary to the inhomogeneous case, such integers may not exist even if we let Q be arbitrarily large. For example, if all the α_i's are multiples of $\frac{1}{2}$, and $\beta_i = \frac{1}{3}$, then $q\alpha_i - p_i$ is also a multiple of $\frac{1}{2}$ and hence

$$|q\alpha_i - p_i - \beta_i| \geq \frac{1}{6}$$

for any choice of $p_1, \ldots, p_n$ and q. Kronecker gave a general condition for the solvability of this problem (see Cassels (1965)):

(1.1.9) Theorem. *For any $2n$ real numbers $\alpha_1, \ldots, \alpha_n$ and $\beta_1, \ldots, \beta_n$, exactly one of the following two conditions holds:*

(i) *For each $\epsilon > 0$, there exist integers $p_1, \ldots, p_n$ and q such that $q > 0$ and*

$$|q\alpha_i - p_i - \beta_i| \leq \epsilon .$$

(ii) *There exist integers $u_1, \ldots, u_n$ such that $u_1\alpha_1 + \ldots + u_n\alpha_n$ is an integer but $u_1\beta_1 + \ldots + u_n\beta_n$ is not.*

Let us remark that if $\alpha_1, \ldots, \alpha_n$, $\beta_1, \ldots, \beta_n$ are rational (which is the case interesting from an algorithmic point of view), then (i) is equivalent to:

(i′) *There exist integers $p_1, \ldots, p_n$ and q such that*

$$q\alpha_i - p_i = \beta_i .$$

So in this case, the inhomogeneous diophantine approximation problem without a bound on q is equivalent to the solution of a special linear diophantine equation. This can be done in polynomial time (Frumkin (1976), Voytakov and Frumkin (1976), von zur Gathen and Sieveking (1976)). But of course we are interested in the problem with an upper bound on q. To find the least common denominator q for a given ϵ is NP–hard (van Emde-Boas (1981)), but we shall describe a method of Babai (1985), which shows that a denominator q

at most 3^n times the smallest, yielding an error at most $3^n\epsilon$, can be found in polynomial time.

1.2. Lattices, bases, and the reduction problem.

A classical tool in the study of diophantine approximation problems is the "geometry of numbers", developed by Minkowski around the turn of the century. This approach involves representing our problems as geometrical problems concerning lattices of points. While the geometric arguments can of course be "translated" into arithmetic (and they have to be if e.g. an algorithm is to be carried out), the geometric insight gained by such representations is a powerful aid to thinking about both classical and algorithmic problems.

Let $a_1, \ldots, a_n \in \mathbb{Q}^n$ be linearly independent vectors; let A be the $n \times n$ matrix $A = (a_1, \ldots, a_n)$. Define the *lattice generated by* $a_1, \ldots, a_n$ as

$$L = L(A) = \mathbb{Z}a_1 + \ldots + \mathbb{Z}a_n = \{\lambda_1 a_1 + \ldots + \lambda_n a_n : \lambda_i \in \mathbb{Z}\} .$$

We say that $a_1, \ldots, a_n$ is a *basis* of L . The same lattice $\mathcal{L}$ may have many bases, but at least they all have the same determinant (up to its sign). So we define

$$\det \mathcal{L} = |\det A|$$

where $A = (a_1, \ldots, a_n)$ is any basis of $\mathcal{L}$. (One may discard the hypothesis here that the number of vectors a_i is as large as the dimension, and study non–full–dimensional lattices. Most of what follows could be extended to this more general case by restricting our attention to the linear subspace generated by $\mathcal{L}$.)

Every lattice $\mathcal{L}$ has a *dual lattice*, called $\mathcal{L}^*$, which is defined by

$$\mathcal{L}^* = \{x \in \mathbb{R}^n : y^T x \in \mathbb{Z} \text{ for every } y \in \mathcal{L}\} .$$

The dual of the standard lattice $\mathbb{Z}^n$ is itself, and the dual of a rational lattice is rational. If $(b_1, \ldots, b_n)$ is a basis of $\mathcal{L}$ then the vectors $c_1, \ldots, c_n$ defined by

$$c_i^T b_j = \begin{cases} 1 & \text{if } i = j , \\ 0 & \text{if } i \neq j \end{cases}$$

form a basis for the dual lattice, called the *dual basis* of $(b_1, \ldots, b_n)$. The determinant of the dual lattice is $\det \mathcal{L}^* = 1/\det\mathcal{L}$.

Geometrically, the determinant of a lattice is the common volume of those parallelohedra whose vertices are lattice points and which contain no other lattice point; equivalently, of those parallelohedra spanned by bases. Hence the following inequality, called *Hadamard's inequality*, is natural from a geometric point of view:

$$||a_1|| \ldots ||a_n|| \geq \det \mathcal{L}. \qquad (1.2.1)$$

(Here $||a||$ denotes the euclidean norm of the vector a. We shall also need the ℓ_1–norm $||a||_1$ and the ℓ_∞–norm $||a||_\infty$.)

Hermite proved that every lattice $\mathcal{L}$ has a basis $(b_1, \ldots, b_n)$ such that

$$||b_1|| \ldots ||b_n|| \leq c_n \det \mathcal{L} ,$$

where c_n is a constant depending only on n. It is known that the best value of c_n is less than n^n.

Note that if $b_1, \ldots, b_n$ is an orthogonal basis then $||b_1|| \ldots ||b_n|| = \det \mathcal{L}$. Hence the ratio $||b_1|| \ldots ||b_n|| / \det \mathcal{L}$ is also called the *orthogonality defect* of the basis.

Let us remark that if we choose n linearly independent vectors $c_1, \ldots, c_n$ in $\mathcal{L}$, then they do not in general form a basis, but, contrary to intuition, the product $||c_1|| \ldots ||c_n||$ may be smaller than for any basis. A well–known theorem of Minkowski on successive minima implies that every lattice $\mathcal{L}$ contains n linearly independent vectors $c_1, \ldots, c_n$ with $||c_1|| \ldots ||c_n|| \leq n^{n/2} \det \mathcal{L}$.

The result of Hermite suggests the following algorithmic problem:

(1.2.2) Basis reduction problem.
Given a lattice $\mathcal{L} = \mathcal{L}(A)$, and $C > 0$, find a basis $(b_1, \ldots, b_n)$ of $\mathcal{L}$ such that

$$||b_1|| \ldots ||b_n|| \leq C \det \mathcal{L} .$$

So this problem has a solution if $C \geq n^n$. It is, however, not known how to find such a basis. To find a basis for which $||b_1|| \ldots ||b_n||$ is minimal is NP–hard.

A related problem is the following.

(1.2.3) Short vector problem.
Given a lattice $\mathcal{L} = \mathcal{L}(A)$ and a number $\lambda > 0$, find a vector $b \in \mathcal{L},\ b \neq 0$ such that $||b|| \leq \lambda$.

Hermite showed that if $\lambda \geq \gamma_n \sqrt[n]{\det \mathcal{L}}$ then such a vector b always exists. Let γ_n denote the smallest constant here. It is known that $\gamma_n \leq \sqrt{\frac{n}{e\pi}}$ and $\gamma_n \geq \sqrt{\frac{n}{2e\pi}}$. Note, however, that a lattice $\mathcal{L}$ may contain a vector much shorter than $\sqrt[n]{\det \mathcal{L}}$.

If we could solve the Short Vector Problem in polynomial time, then of course by binary search over λ, we could also find a shortest non–zero vector in the lattice. We shall denote the length of a shortest non–zero vector in the lattice $\mathcal{L}$ by $\lambda(\mathcal{L})$. It is not known whether the Short Vector Problem (or the Shortest Vector Problem) is NP–complete, although I suspect that it is. It is not known either whether the weaker problem of finding a solution if $\lambda = \sqrt{n} \cdot \sqrt[n]{\det \mathcal{L}}$ is NP–hard.

A more "round" result follows from Minkowski's theorem on lattice points in centrally symmetric convex bodies, if we replace the euclidean norm by the ℓ_∞–norm.

(1.2.4) Theorem. *Every lattice $\mathcal{L}$ contains a vector $b \in \mathcal{L}$, $b \neq 0$ with $||b||_\infty \leq \sqrt[n]{\det \mathcal{L}}$.*

(Note that $\gamma_n \leq \sqrt{n}$ follows from this immediately, since $||b|| \leq \sqrt{n} \cdot ||b||_\infty$.)

Since we want to make some comments on it, let us sketch the (standard) proof of this theorem. We may assume without loss of generality that $\det \mathcal{L} = 1$. Let Q denote the cube $Q = \{x \in \mathbb{R}^n : ||x||_\infty \leq \frac{1}{2}\}$, i.e. the unit cube with center 0. Then vol $(Q) = 1$. Consider the cubes $Q + b$, $b \in \mathcal{L}$. Some two of these cubes must intersect; in fact, if they were disjoint then the density of their union would be less than 1, but on the other hand it is exactly one, as follows by a simple counting argument. So there are two vectors $b_1, b_2 \in \mathcal{L}$, $b_1 \neq b_2$ such that $(Q + b_1) \cap (Q + b_2) \neq \emptyset$. But then $||b_1 - b_2||_\infty \leq 1$ and since $b_1 - b_2 \in \mathcal{L}$, we are done.

The above proof, of course, can be turned into a finite algorithm. If $Q + b_1$ intersects $Q + b_2$ then $Q + (b_1 - b_2)$ intersects Q. So it suffices to look for a cube of the form $Q + b$, $b \in \mathcal{L}$, which intersects Q. Moreover, if $\mathcal{L}$ is given by a basis $A = (a_1, \ldots, a_n)$, then every vector $b = \lambda_1 a_1 + \ldots + \lambda_n a_n \in \mathcal{L}$ that needs to be considered here satisfies

$$|\lambda_i| = |\det (a_1, \ldots, a_{i-1}, b, a_{i+1}, \ldots, a_n)| \leq 2^{\langle A \rangle} .$$

This, however, means the consideration of $2^{n \cdot \langle A \rangle}$ vectors, which is a very "badly" exponential number: it is exponential in the input size even if n is fixed, and it is exponential in n even if the entries of A are "very decent", say $\{0, \pm 1\}$–vectors.

We shall see that one can find in polynomial time a vector satisfying Minkowski's bound (indeed, a shortest vector) if n is fixed (Lenstra (1983)), and also that we can find in polynomial time a vector not longer than $2^n \sqrt[n]{\det \mathcal{L}}$ even for varying n. On the other hand, Lagarias (1983) proved that to find a non–zero vector in a lattice $\mathcal{L}$ which is shortest in the ∞–norm is NP–complete.

Let us return to the euclidean norm for the rest of this chapter. Since from now on we shall be concerned only with finding short vectors or short bases that come within a factor of the optimum, where the factor is a function of n, it does not make any essential difference which norm we use.

Any algorithm that produces a lattice vector "almost" as short as the shortest non–zero vector must also (implicitly) show that the lattice does not contain any vector that is much shorter. Let us turn this around and state first a simple lower bound for the length of the shortest non–zero vector in a lattice. The algorithm described below will make use of this bound.

We need a very basic construction from linear algebra.

Let $(b_1, \ldots, b_n)$ be a basis of $\mathbb{R}^n$. Let us denote by $b_i(j)$ the component of b_i orthogonal to $b_1, \ldots, b_{j-1}$. So $b_i(j) = 0$ if $i < j$. We set $b_j^* = b_j(j)$. Then $(b_1^*, \ldots, b_n^*)$ is a basis, which is called the *Gram–Schmidt orthogonalization* of $(b_1, \ldots, b_n)$. (Note that the b_i^* depend also on the order of the original basis vectors.) The vectors $b_1^*, \ldots, b_n^*$ are mutually orthogonal, and they can be computed from $(b_1, \ldots, b_n)$ by the recurrence

$$b_1^* = b_1 ,$$
$$b_i^* = b_i - \sum_{j=1}^{i-1} \frac{b_i^T b_j^*}{||b_j^*||^2} b_j^* \qquad (i = 1, \ldots, n) .$$

We can write this recurrence in the following form:

$$b_i = \sum_{j=1}^{i} \mu_{i,j} b_j^* \quad (i = 1, \ldots, n) \tag{1.2.5}$$

where $\mu_{ii} = 1$. This form shows that for all $1 \le i \le \mu$, $b_1, \ldots, b_i$ span the same subspace as $b_1^*, \ldots, b_i^*$.

It is also easy to see that

$$\det(b_1, \ldots, b_n) = \det(b_1^*, \ldots, b_n^*) = ||b_1^*|| \ldots ||b_n^*|| .$$

We can now state a lower bound on the length of the shortest non–zero vector in $\mathcal{L}$.

(1.2.6) Lemma. *Let $b_1, \ldots, b_n$ be a basis of a lattice $\mathcal{L}$ and let $b_1^*, \ldots, b_n^*$ be its Gram–Schmidt orthogonalization. Then*

$$\lambda(\mathcal{L}) \ge \min \{||b_1^*||, \ldots, ||b_n^*||\} .$$

Proof. Let $b \in \mathcal{L}$, $b \ne 0$. Then we can write

$$b = \sum_{i=1}^{k} \lambda_i b_i ,$$

where $1 \le k \le n$, $\lambda_i \in \mathbb{Z}$ and $\lambda_k \ne 0$ (we have omitted the 0 terms from the end of this sum). Substituting from (1.2.5) we obtain

$$b = \sum_{i=1}^{k} \lambda_i' b_i^* ,$$

where by $\mu_{kk} = 1$, we know that $\lambda_k' = \lambda_k$ is a non–zero integer. Hence

$$||b||^2 = \sum_{i=1}^{k} \left|\lambda_i'\right|^2 ||b_i^*||^2 \ge |\lambda_k|^2 \, ||b_k^*||^2 \ge ||b_k^*||^2 ,$$

which proves the lemma. □

The following observation shows another connection between the length of basis vectors and the length of their orthogonalizations. Let $(b_1, \ldots, b_n)$ be any basis of a lattice $\mathcal{L}$ and $(b_1^*, \ldots, b_n^*)$ its orthogonalization. Write, as before,

$$b_i = \sum_{j=1}^{i} \mu_{ij} b_j^* \quad (i = 1, \ldots, n) .$$

Then there is another basis $(\bar{b}_1, \ldots, \bar{b}_n)$ with the same orthogonalization $(b_1^*, \ldots, b_n^*)$ such that if we write

$$\bar{b}_i = \sum_{j=1}^{i} \bar{\mu}_{ij} b_j^* \quad (i = 1, \ldots, n)$$

then $|\bar{\mu}_{ij}| \le \frac{1}{2}$ for $1 \le j < i \le n$. A basis $(\bar{b}_1, \ldots, \bar{b}_n)$ with this property will be called *weakly reduced.* In fact, a weakly reduced basis is easily constructed by the following procedure. Suppose that $(b_1, \ldots, b_n)$ is not weakly reduced, then there are indices i and j, $1 \le j < i \le n$ such that $|\mu_{ij}| > \frac{1}{2}$. Choose such a pair with j as large as possible, and let m denote the integer nearest to μ_{ij} . Then $(b_1, \ldots, b_{i-1}, b_i - mb_j, b_{i+1}, \ldots, b_n)$ is a basis of $\mathcal{L}$, it has the same orthogonalization $(b_1^*, \ldots, b_n^*)$ as the original basis, and expressing the basis vectors as linear combinations of the orthogonalized basis vectors only the ith expression changes:

$$b_i - mb_j = \sum_{t=1}^{i} \mu'_{it} b_t^* ,$$

where $\mu'_{it} = \mu_{it}$ for $j < t \le i$, and $\mu'_{it} = \mu_{it} - m\mu_{jt}$ for $1 \le t \le j$. So in particular we still have $\left|\mu'_{it}\right| \le \frac{1}{2}$ for $j < y \le i$, and in addition we now have $\left|\mu'_{ij}\right| = |\mu_{ij} - m| \le \frac{1}{2}$. Repeating this at most $\binom{n}{2}$ times, we obtain a weakly reduced basis.

We now turn to our main basis reduction algorithm for finding a short vector (in fact, a basis consisting of reasonably short vectors). The algorithm will also produce a "certificate", via Lemma (1.2.6), that no substantially shorter vector exists. So we want to find a basis which consists of "short" vectors on the one hand, and whose Gram–Schmidt orthogonalization consists of reasonably long vectors on the other. The following short discussion will be useful for illuminating the situation. Clearly

(1.2.7) $$||b_i^*|| \le ||b_i||$$

(since b_i^* is a projection of b_i) , and

(1.2.8) $$||b_1^*|| \ldots ||b_n^*|| = \det \mathcal{L}$$

is fixed. Typically, the short vectors among $b_1^*, \ldots, b_n^*$ will be at the end of the sequence. So we shall try to make the sequence $\|b_1^*\|, \ldots, \|b_n^*\|$ lexicographically as small as possible. Also, we shall try to make the gap in (1.2.7) as small as possible. These arguments motivate the following definition.

Definition. A basis $(b_1, \ldots, b_n)$ of a lattice $\mathcal{L}$ is called *reduced* if it is weakly reduced, and

$$\|b_i(i)\|^2 \leq \frac{4}{3}\|b_{i+1}(i)\|^2 \quad \text{for } 1 \leq i < n \ . \tag{1.2.9}$$

(Without the coefficient $\frac{4}{3}$, condition (1.2.9) would mean that if b_i and b_{i+1} were interchanged, the orthogonalized sequence $\|b_1^*\|, \ldots, \|b_n^*\|$ would not decrease lexicographically. The coefficient $\frac{4}{3}$ is there for technical reasons, to ensure faster convergence. It could be replaced by any number larger than 1 but less than $\frac{3}{2}$.)

The main properties of this notion of reducedness are summarized in the following two theorems (A. K. Lenstra, H. W. Lenstra, Jr. and L. Lovász (1982)):

(1.2.10) Theorem. *Let $(b_1, \ldots, b_n)$ be a reduced basis of the lattice $\mathcal{L}$. Then*

(i) $\|b_1\| \leq 2^{(n-1)/2}\lambda(\mathcal{L})$;

(ii) $\|b_1\| \leq 2^{(n-1)/4}\sqrt[n]{\det \mathcal{L}}$;

(iii) $\|b_1\| \ldots \|b_n\| \leq 2^{\frac{1}{2}\binom{n}{2}} \det \mathcal{L}$.

(1.2.11) Theorem. *Given a non-singular matrix $A \in \mathbb{Q}^{n\times n}$, a reduced basis in $\mathcal{L} = \mathcal{L}(A)$ can be found in polynomial time.*

Let us sketch the proofs of these results. Without loss of generality we may assume that A has integral entries. Let $(b_1, \ldots, b_n)$ be a reduced basis. Then by (1.2.9) and weak reducedness,

$$\begin{aligned}
\|b_i^*\|^2 = \|b_i(i)\|^2 &\leq \frac{4}{3}\|b_{i+1}(i)\|^2 \\
&= \frac{4}{3}\|b_{i+1}^* + \mu_{i+1,i}b_i^*\|^2 \\
&= \frac{4}{3}\|b_{i+1}^*\|^2 + \frac{4}{3}\mu_{i+1,i}^2\|b_i^*\|^2 \\
&\leq \frac{4}{3}\|b_{i+1}^*\|^2 + \frac{1}{3}\|b_i^*\|^2
\end{aligned}$$

and hence

$$\|b_{i+1}^*\|^2 \geq \frac{1}{2}\|b_i^*\|^2 \ .$$

So by induction,

$$\|b_i^*\|^2 \geq 2^{1-i}\|b_1^*\|^2 = 2^{1-i}\|b_1\|^2 \tag{1.2.12}$$

and thus by Lemma (1.2.6),

$$\begin{aligned}||b_1||^2 &\le \min_i\{2^{i-1}||b_i^*||^2\} \\ &\le 2^{n-1}\min_i||b_i^*||^2 \le 2^{n-1}\lambda(\mathcal{L})^2 .\end{aligned}$$

This proves (i). We also have by (1.2.12) that

$$||b_1||^{2n} \le 2^{n(n-1)/2}\prod_{i=1}^{n}||b_i^*||^2 = 2^{\binom{n}{2}}(\det \mathcal{L})^2 ,$$

whence (ii) follows. Finally, we can estimate $||b_i||$ as follows:

$$\begin{aligned}||b_i||^2 &= \sum_{j=1}^{i}\mu_{ij}^2||b_j^*||^2 \le \sum_{j=1}^{i-1}\frac{1}{4}||b_j^*||^2 + ||b_i^*||^2 \\ &\le \left(1+\frac{1}{4}(2+\ldots+2^{i-1})\right)||b_i^*||^2 \le 2^{i-1}||b_i^*||^2\end{aligned}$$

and hence

$$\prod_{i=1}^{n}||b_i||^2 \le 2^{\binom{n}{2}}\prod_{i=1}^{n}||b_i^*||^2 = 2^{\binom{n}{2}}\det \mathcal{L} .$$

This proves (iii).

The algorithm which is claimed in Theorem (1.2.11) is quite obvious. We maintain a basis $(b_1,\ldots,b_n)$, which we initialize by $(b_1,\ldots,b_n)=(a_1,\ldots,a_n)$. Two kinds of steps alternate: in Step I, we achieve (1.2.8), i.e. we make the basis weakly reduced by the simple pivoting procedure described above. In Step II, we look for an i violating (1.2.9), and then interchange b_i and b_{i+1} .

It is obvious that if we get stuck, i.e. neither Step I nor Step II can be carried out, then we have a reduced basis. What is not so clear is that the algorithm terminates, and in fact in polynomial time.

To this end, let us consider the following quantity:

$$D(b_1,\ldots,b_n) = \prod_{i=1}^{n}||b_i^*||^{n-i} .$$

It is obvious that Step I does not change $D(b_1,\ldots,b_n)$, and it is easy to verify that Step II reduces it by a factor of $2/\sqrt{3}$ or more (this is where the factor $\frac{4}{3}$ in the definition of reducedness plays its role!). Now $D(a_1,\ldots,a_n) \le (\max_i||a_i||)^{\binom{n}{2}}$. On the other hand, let $(b_1,\ldots,b_n)$ be any basis of $\mathcal{L}$, then for any i ,

$$||b_1^*||^2\ldots||b_k^*||^2 = \det(b_i^Tb_j)_{i,j=1}^k \ge 1$$

and hence $D(b_1,\ldots,b_n) \geq 1$. So after p steps of the algorithm, we have

$$1 \leq D(b_1,\ldots,b_n) \leq \left(\frac{\sqrt{3}}{2}\right)^p D(a_1,\ldots,a_n) \leq \left(\frac{\sqrt{3}}{2}\right)^p (\max_i ||a_i||)^{\binom{n}{2}}$$

and hence

$$p \leq 6\binom{n}{2} \max_i \log||a_i|| \leq 6\binom{n}{2}\langle A\rangle .$$

This proves that the procedure terminates after a polynomial number of steps. Thus the whole reduction procedure requires only a polynomial number of arithmetic operations. It still remains to be shown that these arithmetic operations have to be performed on numbers whose input size does not exceed a polynomial of the input size of the problem. This can be done by similar, albeit slightly more tedious, computations. (Note that in the proofs of (i) and (ii), the condition of weak reducedness was only partially used: only $|\mu_{i+1,\,i}| \leq \frac{1}{2}$ was needed. So a vector with property (i) could be found in a smaller number of arithmetic operations, achieving only $|\mu_{i+1,i}| \leq \frac{1}{2}$ in Step I. However, it seems that to guarantee that the numbers occurring in the procedure do not grow too big, one does need the full strength of weak reducedness.)

We close this section with a discussion of some algorithmic implications among the main problems studied here, and with some improvements on our results. The exponential factors $2^{(n-1)/4}$ etc. in Theorem (1.2.10) are rather bad and an obvious question is can they be replaced by a more decent function, at the expense of another, maybe more complicated but still polynomial reduction algorithm. By replacing the "technical factor" $\frac{4}{3}$ in (1.2.9) by $1+\epsilon$, where ϵ is positive but very small, we could replace the base number 2 in the factor $2^{(n-1)/4}$ by any number $c > 2/\sqrt{3}$.

Let us also remark that already this algorithm can be used to find a *shortest* lattice vector in polynomial time if the dimension n of the space is fixed (of course, the polynomial in the time bound will depend on n) . For, if $(b_1,\ldots,b_n)$ is a reduced basis and $z = \sum_{i=1}^n \lambda_i b_i \in \mathbb{R}^n$, then it is not difficult to show that

$$|\lambda_i| < 3^n \cdot ||z||/||b_1|| .$$

Since we are only interested in vectors z with $||z|| \leq ||b_1||$, this means that it suffices to check all vectors $z = \sum_{i=1}^n \lambda_i b_i$ such that $\lambda_i \in \{-3^n,\ldots,3^n\}$. This is a large number, but for n fixed, it is only a constant.

The fact that in bounded dimension we can find a shortest vector in polynomial time (which of course also follows from H. W. Lenstra's integer programming algorithm in bounded dimension, see Section 2.4), was used by Schnorr (1985) to obtain a more substantial improvement over the basis reduction algorithm described above (at least in a theoretical sense). He proved that for every fixed $\epsilon > 0$, we can find a basis $(b_1,\ldots,b_n)$ in polynomial time such that

$$||b_1|| \leq (1+\epsilon)^n \lambda(\mathcal{L})$$

and

$$||b_1|| \ldots ||b_n|| \le (1+\epsilon)^{n^2} \det \mathcal{L} .$$

(The polynomial of course depends on ϵ .)

Schnorr's method can be described as follows. Let $(b_1, \ldots, b_n)$ be a weakly reduced basis. Condition (1.2.9), in the definition of reducedness, if we disregard the "technical factor" $\frac{4}{3}$, says that $b_i(i)$ is not longer than $b_{i+1}(i)$. However, using weak reducedness it is easy to see that this is the same as saying that $b_i(i)$ is not longer than any vector in the lattice generated by $b_i(i)$ and $b_{i+1}(i)$. This observation suggests the idea of fixing any integer $k \ge 1$ and saying that the lattice $\mathcal{L}$ is *k–reduced* if it is weakly reduced and, for all $1 \le i \le n$,

$$||b_i(i)|| \le \frac{4}{3}\lambda(\mathcal{L}(b_i(i), b_{i+1}(i), \ldots, b_{i+k}(i)))$$

(if $i+k > n$, then we disregard the undefined vectors among the generators on the right hand side).

Now Theorem (1.2.10) can be generalized to k–reduced bases as follows:

(1.2.13) Theorem. *Let $(b_1, \ldots, b_n)$ be a k-reduced basis of the lattice $\mathcal{L}$. Then*

(i) $||b_1|| \le a_k \cdot (k+1)^{n/k}\lambda(\mathcal{L})$,

(ii) $||b_1|| \le a_k (k+1)^{n/(2k)} \sqrt[n]{\det \mathcal{L}}$,

(iii) $||b_1|| \ldots ||b_n|| \le a_k^n (k+1)^{n^2/(2k)} \det \mathcal{L}$,

where a_k is a constant depending on k only.

Note that $(k+1)^{1/k} \to 1$ as $k \to \infty$. So if we choose k sufficiently large, then e.g. the coefficient of $\lambda(\mathcal{L})$ in (i) is less than $(1+\epsilon)^n$ for large n .

We only sketch the basic idea of the proof. Let $1 \le t \le k$, and let $\mathcal{L}_{i,t}$ denote the lattice generated by $b_i(i), b_{i+1}(i), \ldots, b_{i+t}(i)$. Then $\det \mathcal{L}_{i,t} = ||b_i^*|| \ldots ||b_{i+t}^*||$, and so by Hermite's theorem,

$$\lambda(\mathcal{L}_{i,t}) \le \gamma_{t+1}^{1/2}(||b_i^*|| \ldots ||b_{i+t}^*||)^{1/(t+1)} .$$

Since $b_i^* = b_i(i)$ is an almost shortest vector in this lattice, we have

$$\begin{aligned} ||b_i^*|| &\le \frac{4}{3}\gamma_{t+1}^{1/2}(||b_i^*|| \ldots ||b_{i+t}^*||)^{1/(t+1)} \\ &< \sqrt{t+1}(||b_i^*|| \ldots ||b_{i+t}^*)^{1/(t+1)} \end{aligned}$$

or

$$(1.2.14) \qquad ||b_i^*||^t \le (t+1)^{(t+1)/2}||b_{i+1}^*|| \ldots ||b_{i+t}^*||.$$

This inequality tells us that the numbers $||b_i^*||$ do not decrease too fast. In fact, one can derive from (1.2.14) by induction on n that

$$||b_1|| = ||b_1^*|| \leq (k+1)^{n/k} \cdot a_k \cdot \min_k ||b_k^*|| ,$$

where a_k is a constant depending only on k. This proves (i). The other two assertions in the theorem follow by similar considerations.

The next thing to show would be that a k-reduced basis can be found in polynomial time. This, however, seems difficult and is not known to hold. Schnorr overcomes this difficulty by changing the notion of k-reducedness in a rather technical manner, whose details are suppressed here.

Even if we give up the idea of finding a shortest vector for variable dimension, it is still interesting to find numbers $a(n), b(n)$ and $c(n)$ as small as possible so that the following tasks can be solved in polynomial time:

(1.2.15) Find a vector $b \in \mathcal{L}$, $b \neq 0$ with $||b|| \leq a(n) \cdot \lambda(\mathcal{L})$.

(1.2.16) Find a vector $b \in \mathcal{L}$, $b \neq 0$ with $||b|| \leq b(n) \sqrt[n]{\det \mathcal{L}}$.

(1.2.17) Find a basis $b_1, \ldots, b_n$ of $\mathcal{L}$ such that $||b_1|| \ldots ||b_n|| \leq c(n) \det \mathcal{L}$.

It would be particularly desirable to be able to solve the problems in polynomial time if $a(n), b(n)$ and $\sqrt[n]{c(n)}$ are polynomials. We do not know if this is possible; but there are some trivial and less trivial relations among these tasks, which in particular show that the problems if $a(n), b(n)$ and $\sqrt[n]{c(n)}$ can be chosen as polynomials are equivalent.

(1.2.18) *If we can solve* (1.2.15) *in polynomial time with some* $a(n)$, *then we can solve* (1.2.16) *with* $b(n) = \sqrt{n}a(n)$ *trivially.*

(1.2.19) *If we can solve* (1.2.15) *in polynomial time with some* $a(n)$, *then we can solve* (1.2.16) *with* $c(n) = n^n \cdot a(n)^n$.

This is a slight generalization of a result of Schnorr (1985). We choose the basis $b_1, \ldots, b_n$ successively. Let b_1 be an "almost shortest" non-zero vector of $\mathcal{L}$, which we can find in polynomial time by hypothesis. Assuming that $b_1, \ldots, b_i$ have already been selected, let us consider the lattice $\mathcal{L}_{i+1}$ obtained by projecting $\mathcal{L}$ on the orthogonal complement of $b_1, \ldots, b_i$ and let b_{i+1}^* be an "almost shortest" vector in $\mathcal{L}_{i+1}$, supplied by our oracle. Let, further, b_{i+1} be any vector in $\mathcal{L}$ whose image in $\mathcal{L}_{i+1}$ is b_{i+1}^* . Then $b_1, \ldots, b_n$ is a basis of $\mathcal{L}$ whose Gram-Schmidt orthogonalization is just $b_1^* = b_1, \ldots, b_n^*$. We may clearly assume that $b_1, \ldots, b_n$ is weakly reduced.

Now we claim that

$$||b_1|| \ldots ||b_n|| \leq n^n \cdot a(n)^n \cdot \det \mathcal{L} .$$

Let $c_1, \ldots, c_n$ be any n linearly independent vectors in $\mathcal{L}$ ordered so that $||c_1|| \leq ||c_2|| \leq \ldots \leq ||c_n||$ (they need not form a basis). Then for each $1 \leq i \leq k$, at least one of $c_1, \ldots, c_k$ is not in the span of $b_1, \ldots, b_{k-1}$. So for this c_j , the projection $c_j(k) \in \mathcal{L}_k$ is non-zero and hence by the choice of b_k^* ,

$$||b_k^*|| \leq a(n)\lambda(\mathcal{L}_k) \leq a(n)||c_j(k)|| \leq a(n)||c_j|| \leq a(n)||c_k|| .$$

So by weak reducedness,

$$||b_k||^2 = \sum_{i=1}^{k} \mu_{ki}^2 ||b_i^*||^2 \le \sum_{i=1}^{k} a(n)^2 ||c_i||^2 \le ka(n)^2 ||c_k||^2$$

and so

$$||b_1|| \dots ||b_{\hat{n}}|| \le \sqrt{n!} a(n)^n ||c_1|| \dots ||c_{\hat{n}}|| \ .$$

It follows from Minkowski's theorem on successive minima that $c_1, \dots, c_{\hat{n}}$ can be chosen so that

$$||c_1|| \dots ||c_{\hat{n}}|| \le \gamma_n^{n/2} \text{ det } \mathcal{L} < n^{n/2} \text{ det } \mathcal{L} \ .$$

So

$$||b_1|| \dots ||b_n|| < n^n a(n)^n \text{ det } \mathcal{L} \ .$$

(1.2.20) *If we can solve* (1.2.17) *in polynomial time with some* $c(n)$, *then we can solve* (1.2.16) *with* $b(n) = \sqrt[n]{c(n)}$, just by choosing the shortest member of the basis.

(1.2.21) Perhaps the most interesting is the observation, based on an idea of H. W. Lenstra and C. P. Schnorr (1984), that *if we can solve* (1.2.16) *in polynomial time with some* $b(n)$, *then we can solve* (1.2.15) *with* $a(n) = b(n)^2$ *in polynomial time.*

In fact, a slightly weaker hypothesis will do. If we choose a vector $c \in \mathcal{L}$ and also a vector $d \in \mathcal{L}^*$ such that $c, d \ne 0$ and

$$||c|| \le b(n) \cdot \text{ det } \mathcal{L}, \qquad ||d|| \le b(n) \cdot \text{ det } \mathcal{L}^* \ ,$$

then

$$||c|| \cdot ||d|| \le b(n)^2 \ .$$

In what follows, we shall use only the consequence that we can find such a pair of vectors in every lattice and its dual in polynomial time.

Let $c_1 \in \mathcal{L}$ and $d_1 \in \mathcal{L}^*$ such that $||c_1|| \cdot ||d_1|| \le b(n)^2$. Assume that we have already chosen non-zero vectors $c_1, \dots, c_k \in \mathcal{L}$ and $d_1, \dots, d_k \in \mathbb{R}^n$ such that $d_1, \dots, d_k$ are mutually orthogonal. Let us consider the lattice $\overline{\mathcal{L}}_k = \{x \in \mathcal{L} : d_1^T x = \dots = d_k^T x = 0\}$, and choose $c_{k+1} \in \overline{\mathcal{L}}_k$ and $d_{k+1} \in \overline{\mathcal{L}}_k^*$ such that $||c_{k+1}|| \cdot ||d_{k+1}|| \le b(n-k)^2 \le b(n)^2$. Since $\overline{\mathcal{L}}_k \subseteq \mathcal{L}$, we have $c_{k+1} \in \mathcal{L}$; on the other hand, $\overline{\mathcal{L}}_k^*$ is not in general a sublattice of $\mathcal{L}^*$, so in general $d_{k+1} \notin \mathcal{L}^*$. But still d_{k+1} belongs to the linear hull of $\overline{\mathcal{L}}_k$, i.e. $d_1^T d_{k+1} = \dots = d_k^T d_{k+1} = 0$. We go on until we have chosen $c_1, \dots, c_n$ and $d_1, \dots, d_n$.

Let c be the shortest vector among $c_1, \dots, c_n$. Then

$$\begin{aligned} ||c|| &= \min(||c_1||, \dots, ||c_n||) \\ &\le b(n)^2 \min \left(\frac{1}{||d_1||}, \dots, \frac{1}{||d_n||} \right) \ . \end{aligned}$$

To complete the argument, it suffices to note that it is easy to construct a basis in $\mathcal{L}$ whose Gram–Schmidt orthogonalization is just $(d_n/||d_n||^2, \ldots, d_1/||d_1||^2)$, and so by Lemma (1.2.6),

$$\lambda(\mathcal{L}) \geq \min\left(\frac{1}{||d_1||}, \ldots, \frac{1}{||d_n||}\right) .$$

So

$$||c|| \leq b(n)^2 \lambda(\mathcal{L})$$

and thus c is a solution of (1.2.15) with $a(n) = b(n)^2$.

Remark. Lenstra and Schnorr have used this argument to show that the lower bound in Lemma (1.2.6) is not too far from $\lambda(\mathcal{L})$: there exists a basis $(b_1, \ldots, b_n)$ in any lattice $\mathcal{L}$ such that

$$\min(||b_1^*||, \ldots, ||b_n^*||) \geq \frac{1}{n}\lambda(H) .$$

Let b be a shortest non–zero vector in the lattice $\mathcal{L}$. We may not be able to prove in polynomial time that b is shortest, but we can prove in polynomial time that b is "almost shortest" in the sense that no non–zero lattice vector is shorter than $||b||/n$. It would be most desirable to design a polynomial–time algorithm that would either conclude that no non–zero lattice vector is shorter than $||b||/n$, or find a vector shorter than b.

Finally, let us address a related problem. Let $\mathcal{L}$ be any lattice in $\mathbb{R}^n$ and $y \in \mathbb{R}^n$. Suppose that we want to find a lattice vector which is nearest to y. Let $d(\mathcal{L}, y)$ denote the minimum distance of y from the vectors in $\mathcal{L}$ (it is obvious that this minimum is attained). To find $d(\mathcal{L}, y)$ is NP–hard (van Emde–Boas (1981)). However, we describe an algorithm that determines this number up to a factor of $(3/\sqrt{2})^n$ in polynomial time, and also finds a vector $b \in \mathcal{L}$ such that $||b - y|| \leq (3/\sqrt{2})^n\ d(\mathcal{L}, y)$.

Similarly, as in the shortest lattice vector case, we may observe that any such algorithm must provide an (implicit) proof of the fact that $d(\mathcal{L}, y)$ is not smaller than $2^{-n}||b - y||$. One obvious bound is the following. Let $w \in \mathcal{L}^*$, then

$$\frac{1}{||w||} d(\mathbb{Z}, w^T y) \leq d(\mathcal{L}, y)$$

(note that $d(\mathbb{Z}, w^T y) = \min \{w^T y - \lfloor w^T y \rfloor,\ \lceil w^T y \rceil - w^T y\}$ and so is easily computable). Khintchine (1948) (see also Cassels (1971, p. 318)) proved that this lower bound is not "too bad" in the sense that there exists a number μ_n such that

$$\mu_n \max_{w \in \mathcal{L}^*} \frac{1}{||w||} d(\mathbb{Z}, w^T y) \geq d(\mathcal{L}, y) .$$

We shall see that $\mu_n = (3/\sqrt{2})^n$ suffices.

To find a vector of $\mathcal{L}$ "close" to y, and at the same time a vector $w \in \mathcal{L}^*$ such that $d(\mathbb{Z}, w^T y)/||w||$ is "big", we describe an algorithm due to Babai (1985).

(1.2.22) Theorem. *Given a non–singular matrix $A \in \mathbb{Q}^{n \times n}$ and a vector $y \in \mathbb{Q}^n$, we can find in polynomial time a vector $b \in \mathcal{L}(A)$ and another vector $w \in \mathcal{L}^*(A)$ such that*

$$||y - b|| \leq \frac{(3/\sqrt{2})^n}{||w||} d(\mathbb{Z}, w^T y) .$$

Hence also

$$||y - b|| \leq (3/\sqrt{2})^n d(\mathcal{L}, y) .$$

Proof. Let $(b_1, \ldots, b_n)$ be any reduced basis of $\mathcal{L} = \mathcal{L}(A)$. We use the notation (1.2.5).

Let us write

$$y = \sum_{i=1}^{n} \lambda_i b_i .$$

Let m_i be the integer next to λ_i and set $\overline{\lambda}_i = \lambda_i - m_i$. Now $b = \sum_{i=1}^n m_i b_i \in \mathcal{L}$ and we shall show that b is a "reasonably good" approximation of y, i.e. that $y' = y - b$ is "short".

Let $(c_1, \ldots, c_n)$ be the basis of $\mathcal{L}^*$ dual to $(b_1, \ldots, b_n)$. Then from

$$y' = \sum_{i=1}^{n} \overline{\lambda}_i b_i$$

we get that $\overline{\lambda}_i = c_i^T y'$ and since $|\overline{\lambda}_i| \leq \frac{1}{2}$, we have that $d(\mathbb{Z}, c_i^T y') = |\overline{\lambda}_i|$. So

$$\begin{aligned} ||y'|| &\leq \sum_{i=1}^{n} |\overline{\lambda}_i| \cdot ||b_i|| = \sum_{i=1}^{n} d(\mathbb{Z}, c_i^T y') \cdot ||b_i|| \\ &\leq \max_i \left\{ \frac{1}{||c_i||} d(\mathbb{Z}, c_i^T y') \right\} \cdot \sum_i ||c_i|| \cdot ||b_i|| . \end{aligned}$$

It remains to estimate the last sum. We know that $||b_i|| \leq 2^{i-1/2} ||b_i^*||$. On the other hand, we can write

$$c_i = \sum_{k=i}^{n} \nu_{ik} b_k^* / ||b_k^*||^2$$

(since $b_n^*/||b_n||^2, \ldots, b_1^*/||b_1^*||^2$ is the Gram–Schmidt orthogonalization of the basis $(c_n, c_{n-1}, \ldots, c_1)$) , and hence

$$\begin{aligned} ||c_i||^2 \cdot ||b_i^*||^2 &= \sum_{k=i}^{n} \nu_{ik}^2 ||b_i^*||^2 \; ||b_k^*||^2 \\ &\le \sum_{k=i}^{n} \nu_{ik}^2 \cdot 2^{k-i} \\ &\le 2^{n-i} \cdot \sum_{k=i}^{n} \nu_{ik}^2 \; . \end{aligned}$$

So

$$\begin{aligned} \sum_{i=1}^{n} ||c_i|| \; ||b_i|| &\le \sum_{i=1}^{n} 2^{i-1/2} ||c_i|| \; ||b_i^*|| \\ &\le 2^{n-1/2} \sum_{i=1}^{n} \left(\sum_{k=i}^{n} \nu_{ik}^2 \right)^{1/2} . \end{aligned}$$

Now the matrix $N = (\nu_{ik})_{i,k=1}^n$ is just the inverse of $M = (\mu_{ik})_{i,k=1}^n$, and hence a routine calculation shows that this sum is at most $2^{n-1/2} \cdot (3/2)^n < 3^n \cdot 2^{-n/2}$. Hence the theorem follows. □

1.3. Diophantine approximation and rounding.

Let $\alpha_1, \ldots, \alpha_n \in \mathbb{Q}$. The problem of finding a good simultaneous approximation of these numbers can be transformed into a short lattice vector problem as follows. Given $0 < \epsilon < 1$ and $Q > 0$, consider the matrix

$$A = \begin{pmatrix} 1 & & 0 & \alpha_1 \\ & \ddots & & \vdots \\ 0 & & 1 & \alpha_n \\ & & & \epsilon/Q \end{pmatrix} \tag{1.3.1}$$

and the lattice $\mathcal{L}(A)$ generated by its columns. Any vector $b \in \mathcal{L}(A)$ can be written as $b = Ap$, where $p = (p_1, \ldots, p_{n+1})^T \in \mathbb{Z}^{n+1}$. Suppose that $b \neq 0$ but $||b|| \le \epsilon$. Then clearly $p_{n+1} \neq 0$.

We may assume without loss of generality that $p_{n+1} < 0$. Let $q = -p_{n+1}$, then we have

$$|b_i| = |p_i - \alpha_i q| \le \epsilon \qquad (i = 1, \ldots, n)$$

and

$$|b_{n+1}| = \frac{\epsilon}{Q} q \le \epsilon \; ,$$

or

$$q \leq Q .$$

Thus, a "short" vector in $\mathcal{L}(A)$ yields a "good" simultaneous approximation of $\alpha_1, \ldots, \alpha_n$.

To make this argument precise, choose $Q = 2^{n(n+1)/4}\epsilon^{-n}$. We know by Theorems (1.2.10) (ii) and (1.2.11) that we can find, in polynomial time, a vector $b \in \mathcal{L}(A)$ such that $b \neq 0$ and

$$||b|| \leq 2^{n/4}\sqrt{\det\mathcal{L}(A)} = 2^{n/4}(\frac{\epsilon}{Q})^{1/(n+1)} = \epsilon$$

and hence

$$||b||_\infty \leq ||b|| \leq \epsilon .$$

Hence we obtain the following theorem.

(1.3.2) Theorem. *Given* $\alpha_1, \ldots, \alpha_n \in \mathbb{Q}$ *and* $0 < \epsilon < 1$, *we can find in polynomial time integers* $p_1, \ldots, p_n$ *and* q *such that*

$$|p_i - q\alpha_i| \leq \epsilon$$

and

$$0 < q \leq 2^{n(n+1)/4}\epsilon^{-n} . \qquad \square$$

We can also view the Small Integer Combination Problem as a special case of the problem of finding a short lattice vector. To this end, let $\alpha_0 = 1$, $\alpha_1, \ldots, \alpha_n \in \mathbb{Q}$, $0 < \epsilon < 1$ and $Q > 0$. Consider the matrix

$$A' = \begin{pmatrix} \alpha_0 & \alpha_1 & \ldots & \alpha_n \\ & \epsilon/Q & & 0 \\ & & \ddots & \\ & 0 & & \epsilon/Q \end{pmatrix}$$

and the lattice $\mathcal{L}(A')$ generated by its columns. Then any $b' \in \mathcal{L}(A')$ can be written as $b' = A'p'$, where $q = (q_0, \ldots, q_n)^T \in \mathbb{Z}^{n+1}$. Suppose that $b' \neq 0$ but $||b'|| \leq \epsilon$. Then

$$|q_0\alpha_0 + \ldots + q_n\alpha_n| \leq \epsilon$$

and

$$\frac{\epsilon}{Q}|q_i| \leq \epsilon, \quad |q_i| \leq Q \quad (i = 1, \ldots, n) .$$

Applying Theorem (1.2.10) (ii) again, we find that if $\epsilon = 2^{n/4}\sqrt[n+1]{\det\mathcal{L}(A')}$ i.e. if $Q = 2^{(n+1)/4}\epsilon^{-1/n}$ then such a vector b can be found in polynomial time. So we obtain the following.

(1.3.3) Theorem. *Given $\alpha_0 = 1, \alpha_1, \ldots, \alpha_n \in Q$ and $\epsilon > 0$, we can find in polynomial time integers $q_0, q_1, \ldots, q_n \in \mathbb{Z}$, not all 0, such that*

$$|q_0\alpha_0 + q_1\alpha_1 + \ldots + q_n\alpha_n| \leq \epsilon$$

and

$$|q_i| \leq 2^{(n+1)/4}\epsilon^{-1/n} \quad (i = 1, \ldots, n) . \qquad \square$$

Remark. If we consider, instead, the matrix

$$A'' = \begin{pmatrix} \alpha_0 & \alpha_1 & \ldots & \alpha_n \\ \epsilon/Q & & 0 & \\ & & \ddots & \\ 0 & & & \epsilon/Q \end{pmatrix}$$

then we could obtain a version of the last theorem where all the coefficients $q_0, \ldots, q_n$ are uniformly bounded. The details of this are left to the reader as an exercise (note that the lattice generated by A'' is not full–dimensional).

The geometric methods developed in the previous section also apply to non–homogeneous diophantine approximation and yield the following result, which may be viewed as an algorithmic, effective version of Kronecker's Theorem (1.1.9).

(1.3.4) Theorem. *Given $\alpha, \beta \in \mathbb{Q}^n$ and $\epsilon, Q > 0$, we can achieve in polynomial time one of the following:*

(i) *find $p \in \mathbb{Z}^n$ and $q \in \mathbb{Z}$ such that $0 < q \leq Q$ and*

$$||q\alpha - p - \beta||_\infty < \epsilon ;$$

(ii) *find $u \in \mathbb{Z}^n$ such that*

$$\sqrt{n} \cdot (3/2)^n d(\mathbb{Z}, u^T\beta) > Qd(\mathbb{Z}, u^T\alpha) + \epsilon||u||_1 .$$

Remark. Conclusions (i) and (ii) are not mutually exclusive, but is easy to see that if (i) has a solution in p and q then for all $u \in \mathbb{Z}^n$, one has

$$d(\mathbb{Z}, u^T\beta) \leq Qd(\mathbb{Z}, u^T\alpha) + \epsilon||u||_1 .$$

Hence if (ii) has a solution for some ϵ and Q then (i) has no solution with $\epsilon' = \epsilon \cdot n^{-1/2}(3/\sqrt{2})^n$ and $Q' = Qn^{-1/2}(3/\sqrt{2})^n$.

Proof. Consider the matrix (1.3.1) and the vector $y = \binom{\beta}{0} \in Q^{n+1}$. Applying Theorem (1.2.22) we can find, in polynomial time, two vectors $b \in \mathcal{L}(A)$ and $w \in \mathcal{L}^*(A)$ such that

$$||y - b|| \leq \frac{(3/\sqrt{2})^n}{||w||} d(\mathbb{Z}, w^T y) .$$

Now there are two cases to distinguish.

Case 1. $||y-b|| < \epsilon$. Then just as before, we set $b = Ap'$, where $p' = \binom{-p}{q} \in \mathbb{Z}^{n+1}$, $q > 0$. So we have

$$||q\alpha - p - \beta||_\infty \leq ||q\alpha - p - \beta|| < \epsilon$$

and also

$$0 < q < Q \, .$$

Case 2. $||y-b|| \geq \epsilon$. Then we set $w = \binom{u}{v}^T$, where $u \in \mathbb{Q}^n$ and $v \in \mathbb{Q}$. So we have

$$\begin{aligned} d(\mathbb{Z}, w^T y) &= d(\mathbb{Z}, u^T \beta) \geq (3/\sqrt{2})^{-n} \epsilon ||w|| \\ &\geq \frac{1}{\sqrt{n}} (3/\sqrt{2})^{-n} \epsilon ||w||_1 \\ &= \frac{1}{\sqrt{n}} (3/\sqrt{2})^{-n} \epsilon \cdot ||u||_1 + \frac{1}{\sqrt{n}} (3/\sqrt{2})^{n} \epsilon \, |v| \; . \end{aligned} \tag{1.3.5}$$

From $w \in \mathcal{L}^*(A)$ we see that $u \in \mathbb{Z}^n$ and also that $u^T\alpha + v\frac{\epsilon}{Q}$ is an integer. Hence

$$d(\mathbb{Z}, u^T \alpha) \leq \frac{\epsilon}{Q} \, |v| \ ,$$

and hence from (1.3.5) we find that (ii) is achieved. □

A slightly different way of applying the basis reduction algorithm to obtain inhomogeneous diophantine approximation was used by Odlyzko and te Riele (1985). They used it in conjunction with deep methods from analytic number theory and with numerical procedures to disprove Mertens' conjecture, a long–standing open problem in number theory. This conjecture asserted that

$$\left| \sum_{k \leq x} \mu(k) \right| \leq \sqrt{x}$$

for all $x \geq 1$, where μ is the Möbius function. They used a transformation of the problem into an assertion involving the roots of the Riemann zeta function, which they disproved obtaining good non–homogeneous simultaneous approximation for 70 of these roots.

For some further applications of these methods, in particular in cryptography, see Lagarias and Odlyzko (1983).

For our purposes, the main use of the simultaneous diophantine approximation will be that it provides a "rounding" process with very nice properties. Let $y \in \mathbb{Q}^n$ be any vector of "data". Suppose that we want to replace y by a "nearby" vector $\overline{y}$ such that $\langle \overline{y} \rangle$ is "small". If we only want that $||\overline{y} - y||$ be small then the best we can do is to round the entries of y independently of each other. But if we use simultaneous diophantine approximation then much more can be achieved: We can require that all linear inequalities which hold true for y and which have relatively simple coefficients remain valid for $\overline{y}$; even more, all such linear inequalities which "almost" hold for y should be "repaired" and hold true for $\overline{y}$. We state this exactly in the following theorem.

(1.3.6) Theorem. *Given $y \in \mathbb{Q}^n$ with $||y||_\infty = 1$ and $k > n$, we can compute in time polynomial in $\langle y \rangle$ and k a vector $\overline{y} \in \mathbb{Q}^n$ such that*

(i) $\langle \overline{y} \rangle \leq 6kn^2$;

(ii) *for any linear inequality $a^T x \leq \alpha$ with $\langle a \rangle + \langle \alpha \rangle \leq k$, if y "almost satisfies" $a^T x \leq \alpha$ in the sense that $a^T y \leq \alpha + 2^{-4nk}$, then we have $a^T \overline{y} \leq \alpha$.*

Note that (ii) implies that $\overline{y}$ satisfies the inequalities $s_i \leq \overline{y}_i \leq r_i$, where s_i, r_i are the rational numbers with input size at most $k - n - 1$ next to y_i . From the results on continued fractions it follows that $|s_i - r_i| \leq 2^{-(k-n-1)}$ and hence $||y - \overline{y}||_\infty \leq 2^{-(k-n-1)}$.

If y satisfies or "almost " satisfies a linear equation $a^T x = \alpha$ with $\langle a \rangle + \langle \alpha \rangle \leq k$, then applying (ii) to $a^T x \leq \alpha$ and $a^T x \geq \alpha$, we find that $\overline{y}$ will satisfy $a^T \overline{y} = \alpha$. In particular if $\langle y_i \rangle \leq k - n - 1$ for some i , then $\overline{y}_i = y_i$.

Proof. We set $\epsilon = 2^{-2k-1}$ and apply Theorem (1.3.1) to find integers $p_1, \ldots, p_n$, q such that $0 < q < 2^{n(n+1)/4} \epsilon^{-n}$ and $|p_i - q y_i| \leq \epsilon$. We claim that $\overline{y} = (\frac{p_1}{q}, \ldots, \frac{p_n}{q})$ satisfies the conditions. Condition (i) is obvious. To show that (ii) holds, let $a^T x \leq \alpha$ be any linear inequality "almost satisfied" by y (i.e. $a^T y \leq \alpha + 2^{-4nk}$) , and assume that $\langle a \rangle + \langle \alpha \rangle \leq k$. Let T denote the least common denominator of the entries of a and α ; then it is easy to see that $T < 2^k$. Now

$$\begin{aligned} a^T \overline{y} - \alpha &= a^T (y - \overline{y}) + a^T \overline{y} - \alpha \\ &\leq ||a||_1 ||y - \overline{y}||_\infty + 2^{-4nk} \\ &\leq 2^k \cdot \frac{\epsilon}{q} + 2^{-4nk} \\ &\leq \frac{1}{q} \cdot 2^{-k} < \frac{1}{qT} . \end{aligned}$$

But $a^T \overline{y} - \alpha$ is an integral multiple of $\frac{1}{qT}$, and so it follows that $a^T \overline{y} - \alpha \leq 0$. □

The fact that the above–described rounding procedure "corrects" small violations of linear inequalities is very important in many applications, as we shall see. In other cases, however, it is quite undesirable, because it implies that the procedure does not preserve strict inequalities. A. Frank and É. Tardos (1985) have found another rounding procedure which preserves "simple" strict inequalities. Their result is the following:

(1.3.7) Theorem. *Given $y \in \mathbb{Q}^n$ with $||y||_\infty = 1$ and $k > n$, we can compute in time polynomial in $\langle y \rangle$ and k a vector $\tilde{y} \in \mathbb{Q}^n$ such that*

(i) $\langle \tilde{y} \rangle \leq 6kn^4$;

(ii) *for any linear equation $A^T x = \alpha$ with $\langle a \rangle + \langle \alpha \rangle \leq k$, if y satisfies it then $\tilde{y}$ satisfies it;*

(iii) *for any strict linear inequality $a^T x < \alpha$ with $\langle a \rangle + \langle \alpha \rangle \leq k$, if y satisfies it then $\tilde{y}$ satisfies it.*

Perhaps the most interesting special case arises when $k = 2n+1$. Then all inequalities and equations of the form $\sum_{i \in I_1} y_i > \sum_{i \in I_2} y_i$ and $\sum_{i \in I_1} y_i = \sum_{i \in I_2} y_i$ are preserved by the rounding process.

Proof. First, we construct the vector $\overline{y}$ obtained by the rounding process in Theorem (1.3.6). If $y = \overline{y}$ we have nothing to do, so suppose that $y \neq \overline{y}$. Let $y_1 = (y - \overline{y})/||y - \overline{y}||_\infty$. Again, apply the rounding process in Theorem (1.3.6) to y_1, and let $\overline{y}_1$ be the resulting vector. Then let $y_2 = (y_1 - \overline{y}_1)/||y_1 - \overline{y}_1||_\infty$, etc.

First we remark that this procedure terminates in at most n steps. In fact, y has a coordinate which is ± 1 by hypothesis, and then this coordinate is the same in $\overline{y}$. Hence y_1 has at least one coordinate 0. Similarly, y_2 has at least two 0's etc., $y_n = 0$. Thus we have obtained a decomposition

$$
\begin{aligned}
y = \overline{y} &+ ||y - \overline{y}||_\infty \cdot \overline{y}_1 + ||y - \overline{y}||_\infty \cdot ||y_1 - \overline{y}_1||_\infty \cdot \overline{y}_2 \\
&+ ||y - \overline{y}||_\infty \cdot ||y_1 - \overline{y}_1||_\infty \cdot \ldots \cdot ||y_{n-2} - \overline{y}_{n-2}||_\infty \cdot \overline{y}_{n-1} \,.
\end{aligned}
$$

Let $\delta = 2^{-6nk}$ and define

$$\tilde{y} = \overline{y} + \delta \cdot \overline{y}_1 + \delta^2 \overline{y}_2 + \ldots \quad .$$

We claim that $\tilde{y}$ satisfies the requirements in the theorem. Condition (i) is easily checked. To verify (ii), let $a^T x = \alpha$ be a "simple" linear equation satisfied by y. Then we also have $a^T \overline{y} = \alpha$ by the properties of the first rounding, and so $a^T(y - \overline{y}) = 0$. Hence $a^T x = 0$ is a "simple" linear equation satisfied by $y_1 = (y - \overline{y})/||y - \overline{y}||_\infty$ and so again it follows that $a^T \overline{y}_1 = 0$. Similarly we find that $a^T \overline{y}_i = 0$ and so

$$a^T \tilde{y} = a^T \overline{y} + \delta a^T \overline{y}_1 + \ldots = \alpha \,.$$

Finally, consider a strict inequality $a^T x < \alpha$ with $\langle a \rangle + \langle \alpha \rangle \leq k$ satisfied by y. Then it follows by the properties of the first rounding procedure that $\overline{y}$ satisfies the non-strict inequality $ay \leq \alpha$.

Case 1. Assume first that $a^T \overline{y} < \alpha$. Let q denote the least common denominator of the entries of $\overline{y}$ and let T be the least common denominator of the entries of a and α. Then $\alpha = a^T \overline{y}$ is an integral multiple of $\frac{1}{qT}$, and hence

$$\alpha - a^T \overline{y} \geq \frac{1}{qT} \geq 2^{-5nk} \,.$$

So

$$\begin{aligned} a^T\tilde{y} &= a^T\overline{y} + \delta a^T\overline{y}_1 + \delta^2 a^T\overline{y}_2 + \dots \\ &\le \alpha - 2^{-5nk} + \delta||a||_1||\overline{y}_1||_\infty + \delta^2||a||_1||\overline{y}_2||_\infty \\ &= \alpha - 2^{-5nk} + ||a||_1(\delta + \delta^2 + \dots) < \alpha \ , \end{aligned}$$

i.e. the strict inequality is preserved.

Case 2. Assume that $a^T\overline{y} = \alpha$. Consider the numbers $a^T\overline{y}_1$, $a^T\overline{y}_2, \dots$. Not all of these can be 0, since then we would find that $a^Ty = \alpha$, contrary to hypothesis. So there is a first index i such that $a^T\overline{y}_i \ne 0$. Now from $a^T\overline{y} = \alpha$ it follows that $a^T(y - \overline{y}) < 0$ and so $a^Ty_1 < 0$. If $a^T\overline{y}_1 = 0$ then it follows that $a^T\overline{y}_2 < 0$ etc., we find that $a^Ty_i < 0$. So by the properties of rounding, we find that $a^T\overline{y}_i \le 0$, and by the choice of i, $a^T\overline{y}_i < 0$. Hence we can conclude by the same argument as in Case 1 that $a^T\tilde{y} < \alpha$, i.e. that the strict inequality is preserved in this case as well. □

1.4. What is a real number?

It is a little black box, with two slots. If we plug in a (rational) number $\epsilon > 0$ on one side, it gives us back a rational number r on the other (which is meant to be an approximation of the real number α described by the box, with error less than ϵ).

It is important to know that the answers given by the box are reasonable; therefore we require that the box wear a little tag like this:

(1.4.1)

MANUFACTURER'S GUARANTEE:
For any two inputs $\epsilon_1, \epsilon_2 > 0$, the outputs r_1 and r_2 satisfy $|r_1 - r_2| < \epsilon_1 + \epsilon_2$.

It is obvious that if we have such a box (and it works as its manufacturer guarantees), then it does indeed determine a unique real number. Such a black box ("oracle") can now be included in any algorithm as a subroutine. Since we do not know how the box works, we shall count one call on this oracle as one step. If we are interested in polynomial–time computations then, however, an additional difficulty arises: the output of the oracle may be too long, and it might take too much time just to read it. So we shall assume that our black boxes also have the following additional tag:

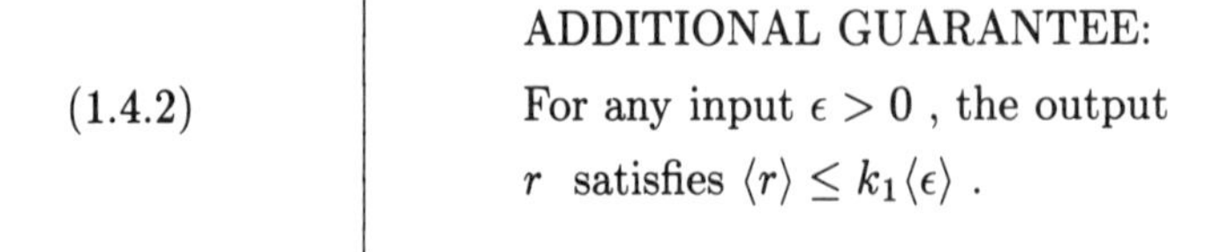
(1.4.2)

ADDITIONAL GUARANTEE:
For any input $\epsilon > 0$, the output r satisfies $\langle r \rangle \le k_1\langle \epsilon \rangle$.

(Here k_1 is a constant explicitly given in unary encoding.) An oracle with guarantees (1.4.1) and (1.4.2) will be called a *real number box*. The number k_1 is the contribution of this box to the input size of any problem in which the box is used as a subroutine.

It is not difficult to assemble a box named $\sqrt{2}$, or a box named π, etc. In fact, one can realize these boxes so that they work in polynomial time. Furthermore, if we have two boxes named "α" and "β" then it is easy to design a box for "$\alpha+\beta$" which satisfies the right kinds of guarantees and which works in polynomial time; similarly for "$\alpha-\beta$" and "$\alpha\cdot\beta$". The situation is more complicated with division. There is no difficulty with designing a box for "α/β", provided we can compute a positive lower bound on $|\beta|$ in polynomial time. But what happens if such a bound cannot be computed? This then means that β cannot be distinguished from 0 in polynomial time. This leads us to the discovery that the equality of two real numbers cannot be determined from the black box description above. At first sight this seems to be a serious handicap of this model, but it in fact reflects a real requirement in numerical analysis: stability. If an algorithm contains, say, a branching depending on the condition $\alpha = \beta$, then it should also be correct if the condition is replaced by $|\alpha - \beta| < \epsilon$ for any sufficiently small ϵ; since if α and β are only approximately known, their exact equality cannot be decided.

We shall not spend much effort here to elaborate this model of real numbers, but it should be remarked that a similar approach has been put forth by many; let us just refer to Turing (1937), Bishop (1967), and Ko (1983). Most of these approaches deal with computability properties of particular real numbers, an interesting question which we cannot touch upon here. Our motivation is not a constructivist philosophy of mathematics either; it is rather to supply a tool to formalize and analyse real number computations. We are also going to point out how this model leads to algorithmic problems which are non–trivial, and whose solutions need much of the machinary developed in this chapter. This may turn out to be no more than a curiosity; but a similar "oracle" model for convex bodies (which is a direct extension of this oracle model) is rather convincingly the "right" framework in which to discuss the main results of the next chapter.

It is natural to ask: why did we take the Cauchy sequence model of real numbers as the basis of our work? Why not the Dedekind cut model? One could do the latter; indeed then the oracle would accept rational numbers r as its input, and answer "my real number is larger/ not larger than r". The two models are not equivalent (cf. Ko (1983)); the "Cauchy sequence oracle" is weaker, and yet sufficient for at least those results which follow.

We shall only address one question concerning real number boxes: how do special classes of real numbers fit in? There is no problem with integers. First, if a is any integer then we may design a box which answers "a" to any query; as output–guarantee we can write "$\langle r\rangle \leq \langle a\rangle \cdot \langle\epsilon\rangle$" on it. Conversely, if we have a real number box which we know represents an integer, then we can recover this integer by plugging in $\epsilon = \frac{1}{2}$ and then rounding the output to the next integer. So integers may be viewed as (and are polynomial time equivalent to) real number boxes with an additional tag:

(1.4.3)

ADDITIONAL GUARANTEE: This number is an integer.

Let us turn to rational numbers. One might think that it suffices to supply those boxes describing a rational number with a tag:

ADDITIONAL GUARANTEE: This number is rational.

Note, however, that this guarantee is meaningless: you can never catch the manufacturer of the box cheating. In fact, after any finite number of calls on any real number oracle, there will always be a rational number satisfying all the previous answers!

The way out is to require a stronger guarantee from the manufacturer; say the following:

(1.4.4)

ADDITIONAL GUARANTEE: This number is rational with input size at most k_2 .

Here k_2 is a constant explicitly given in unary encoding. This value k_2 must be added to the input size of the box, which will thus be $k_1 + k_2$ (k_1 from tag (1.4.2) and k_2 from tag (1.4.4)).

(1.4.5) Theorem. *Rational numbers and real number boxes with guarantee (1.4.4) are polynomial time equivalent.*

Proof. First, given a rational number a , we can easily design a box which answers a to any possible query, and take $k_1 = k_2 = \langle a \rangle$ in the guarantees.

Second, assume that we have a real number box with a correct rationality guarantee (1.4.4). Plug in $\epsilon = 2^{-2k_2-1}$, then we obtain from it a rational number r such that the rational number p/q in question lies in the interval $(r-\epsilon, r+\epsilon)$. We know that $q < 2^{k_2}$. The difference of two rational numbers with denominator less that 2^{k_2} is at least $2^{-2k_2} = 2\epsilon$. So the interval $(r-\epsilon, r+\epsilon)$ contains at most one such number.

So the box, with the given guarantees, determines p/q uniquely. But how can we compute p/q? Since the interval $(r-\epsilon, r+\epsilon)$ contains no other rational number with denominator less than 2^{k_2} , the number p/q is the rational number with denominator less that 2^{k_2} closest to r . In Section 1.1 we have shown that this number can be determined in polynomial time in $\langle 2^{k_2}\rangle + \langle r\rangle \le k_2 + 1 + k_1\langle\epsilon\rangle = k_1 + 3k_2 + 4$. This proves Theorem (1.4.5.). □

We face deeper problems if we go a step further and study *algebraic numbers.* Let us recall from classical algebra that a complex number α is *algebraic* if there is a polynomial $f(x) = a_0 + a_1x + \ldots + a_nx^n$ with integral (or, equivalently, with rational) coefficients such that α is a root of f , i.e. $f(\alpha) = 0$. It is well known that among these polynomials there is one, called the *minimal polynomial* of α , such that all the others are multiples of this. This minimal polynomial is uniquely determined up to multiplication by a constant.

A non–constant polynomial f with rational coefficients is called *irreducible* if it cannot be obtained as the product of two non–constant polynomials with rational coefficients. Every polynomial has an essentially unique factorization into irreducible polynomials. The minimal polynomial of any algebraic number is irreducible.

One way to describe an algebraic number, at least if it is real, is by a real number box with an additional guarantee:

(1.4.6)

ADDITIONAL GUARANTEE:
This number is algebraic and its minimal polynomial has input size $\le k_3$.

If α is not real, we can describe it by a pair of such boxes, one for its real part and one for its imaginary part. But one also would like to think about algebraic numbers as $\sqrt{2}$, $\sqrt[3]{5}$ etc. What does $\sqrt{2}$ mean? It means "the unique root of the polynomial $x^2 - 2$ in the interval $(0, \infty)$". This suggests that we use their minimal polynomials to specify algebraic numbers. It will be worth allowing a little more: we use any polynomial f with rational coefficients which has α as a root to specify α .

A polynomial has in general more than one root, so we also have to specify which of them we mean. If α is real, then this can be done e.g. by specifying an interval (a, b) $(a, b \in \mathbb{Q})$ such that α is the unique root of f in (a, b) . For complex numbers, we can replace this interval by a square in the complex plane.

So a real algebraic number can be encoded as a triple $(f; a, b)$, where $f(x) = a_0 + a_1x + \ldots + a_nx^n$ is a polynomial with rational coefficients and (a, b) is an interval which contains exactly one root of f . The *input size* of such an encoding is

$$\langle f\rangle + \langle a\rangle + \langle b\rangle = \sum_{i=0}^{n}\langle a_i\rangle + \langle a\rangle + \langle b\rangle .$$

The input sizes $\langle a\rangle$ and $\langle b\rangle$ do not matter very much. It is not difficult to see that the roots of f are spread out sufficiently well so that one can always find appropriate a and b such that $\langle a\rangle, \langle b\rangle$ are bounded by a polynomial of $\langle f\rangle$.

On the other hand, it is important that in the definition of $\langle f\rangle$, we consider all terms, even those with $a_i = 0$. Hence in particular $\langle f\rangle > n$.

This encoding of algebraic numbers is not unique; it is, however, not difficult to find a polynomial–time test to see if two triples $\langle f; a, b\rangle$ and $\langle g; c, d\rangle$ describe the same algebraic number. It is somewhat more difficult to compute triples encoding the sum, difference, product and ratio of two algebraic numbers, but it can also be done in polynomial time.

It can be perhaps expected, but it is by no means easy to show, that the descriptions of real algebraic numbers given above are polynomial time equivalent. This can be stated exactly as follows.

(1.4.7) Theorem. (a) *Given an algebraic number triple $(f; a, b)$ and an $\epsilon > 0$, we can compute in polynomial time a rational number r which approximates the root α of f in (a, b) with error less than ϵ .*

(b) *Given a rational number r , a (unary) integer k , and the information that there exists a real algebraic number α whose minimal polynomial has input size at most k and for which $|r - \alpha| < 2^{-4k^3}$, we can compute the minimal polynomial of α in polynomial time.*

Proof. (Sketch.) (a) is easy by binary search over the interval (a, b) . The hard part is (b). To solve this, let $f(x) = a_0 + a_1x + \ldots + a_mx^m$ be the minimal polynomial of α (which we want to determine). We do know that $m \le k$. Let $1 \le n \le k$, and consider the matrix

$$A = \begin{pmatrix} 1 & \alpha & \ldots & \alpha^n \\ 2^{-3k^3} & & & 0 \\ & 2^{-3k^3} & & \\ & & \ddots & \\ 0 & & & 2^{-3k^3} \end{pmatrix}.$$

Let $\mathcal{L}(A)$ be the lattice generated by the columns of A , and let $b \in \mathcal{L}(A)$, $b \neq 0$ be a vector in $\mathcal{L}(A)$ which is "almost shortest" in the sense that $||z|| \ge 2^{-n}||b||$ for every $z \in \mathcal{L}(A)$, $z \neq 0$.

How short is this b ? Suppose first that $n \ge m$. Combining the columns of a with coefficients $a_0, \ldots, a_m$, and $a_{m+1} = \ldots = a_n = 0$, we find that the vector $z = (0 = f(\alpha), 2^{-3k^3}a_0, \ldots, 2^{-3k^3}a_n)$ belongs to $\mathcal{L}(A)$. So in this case

$$||b|| \le 2^n||z|| \le 2^{n-3k^3}\sqrt{\sum a_i^2} .$$

Here $|a_i| \le 2^k$ by the hypothesis on the minimal polynomial of α and so

$$||b|| \le 2^{n+k-3k^3} < 2^{2k-3k^3} .$$

Conversely, suppose that $b \in \mathcal{L}(A)$ is any vector such that

$$||b|| \le 2^{2k-3k^3}.$$

We can write $b = (g(\alpha), 2^{-3k^3}b_0, \ldots, 2^{-3k^3}b_n)^T$, where $g(x) = b_0 + b_1x + \ldots + b_nx^n$ is some polynomial with integral coefficients. Hence in particular

$$|g(\alpha)| \le 2^{2k-3k^3}$$

and

$$|b_i| \le 2^{2k} .$$

We claim that $g(\alpha) = 0$. Let $\alpha = \alpha_1, \alpha_2 \ldots \alpha_m$ be the conjugates of α , i.e. the roots of its minimal polynomial. Then it is known from classical algebra that $a_m^m g(\alpha_1) \ldots g(\alpha_m)$ is an integer. But it follows by routine computation that $|\alpha_i| \le 2^k$ and hence $|g(\alpha_i)| \le 2^{k^2+3k}$. Thus

$$|a_m^m g(\alpha_1) \ldots g(\alpha_m)| \le 2^{km} \cdot 2^{2k-3k^3} \cdot (2^{k^2+3k})^{m-1} < 1 ,$$

which shows that $g(\alpha_1) \ldots g(\alpha_m) = 0$. So $g(\alpha_i) = 0$ for some i . But since $\alpha = \alpha_1, \ldots, \alpha_m$ are conjugates, this implies that $g(\alpha) = 0$.

So we can find the minimal polynomial by the following procedure: for $n = 1, 2, \ldots, k$, we determine an "almost shortest" vector b in the lattice $\mathcal{L}(A)$. If $||b|| > 2^{2k-3k^3}$, then the minimal polynomial of α has larger degree, and we can move on to the next n . If $||b|| < 2^{2k-3k^3}$, then the coefficients $b_0, \ldots, b_n$ in the representation of b as an integer linear combination of the columns of $\mathcal{L}(A)$ yield the minimal polynomial of α .

There is still some cheating in this argument, since we do not know α and hence we cannot work with $\mathcal{L}(A)$. But r is a sufficiently good approximation of α , so that we can replace α by r in the definition of A and the same argument holds. □

Remark. In order that part (a) of Theorem (1.4.7) can really be interpreted as "given a triple, we can design a box", we also need an upper bound on the input size of the minimal polynomial of α . We do know one polynomial f has α as a root, but the input size of the minimal polynomial g of α may be larger than the input size of f. Fortunately it follows from a result of Mignotte (1974) that $\langle g \rangle \le 2n\langle f \rangle$, so this is easily achieved.

We conclude with an application of these considerations. The following result was proved by A. K. Lenstra, H. W. Lenstra and L. Lovász (1982) in a different way. The algorithm given here was sketched in that paper and elaborated on by R. Kannan, A. K. Lenstra and L. Lovász (1984).

(1.4.8) Corollary. *A polynomial with rational coefficients can be factored into irreducible polynomials in polynomial time.*

Proof. Let α be any root of the given polynomial f. For simplicity, assume that α is real (else, we could apply a similar argument to the real and imaginary parts of α). Using Theorem (1.4.7) (a), we can design a real number box description of α. Using part (b) of this same theorem, we can determine the minimal polynomial g of α in polynomial time.

Now if $f = g$ then f is irreducible and we have nothing to prove. If $f \neq g$ then g divides f by the fundamental property of the minimal polynomial, and then we have the decomposition $f = g \cdot (f/g)$. Replacing f by f/g and repeating this argument, we obtain a decomposition of f into irreducible factors. □

CHAPTER 2

How to Round a Convex Body

2.0. Preliminaries: Inputting a set.

In this chapter we shall consider algorithms involving a convex set. Convex sets occur in a large number of algorithmic problems in numerical analysis, linear programming, combinatorics, number theory and other fields. Due to this diversity of their origin, they may be given to us in very many forms: as the solution set of a system of linear inequalities, or as the convex hull of a set of vectors (where this set itself may be given explicitly or implicitly, like the set of incidence vectors of matchings in a graph), or as the epigraph of a function etc. So a problem like "determine the volume of a convex body" is not well posed: the algorithms one chooses will depend on how the set is given.

For lattices, we have always assumed that they were given by a basis. This hypothesis was good enough for the algorithms discussed in the preceding chapter (see, however, Lovász (1985) for a discussion of other possible descriptions of lattices).

For convex sets, however, there is no single most natural description. So we will take the approach of formulating several alternative ways of description (using the notion of oracles) and will prove theorems establishing the algorithmic equivalence of these. Similarly to the ways in which we obtained the results in Section 1.4, we can apply these to explicit special cases and obtain non-trivial algorithms. For example, the polynomial-time solvability of linear programming will occur as such a special case. Many further applications will be contained in Chapter 3.

Let us include here a brief informal discussion of inputting a set.

> We may encode the set as an explicit list of its elements. This only works for finite sets; even then it may be impractical if the set is very large in comparison with the "natural" input size of the problem, e.g. the set of all matchings in a graph.
>
> We may encode a set S by providing a subroutine (oracle) which tells us whether or not a given element of some natural large universe (e.g. $\mathbb{Q}^n$

in the case of a convex set) belongs to S . We shall call such an oracle a *membership oracle.*

It may be that if the oracle answers that $a \notin S$ then there is always a simple reason for this, and we may require the oracle to supply this reason. For closed convex sets in $\mathbb{R}^n$, for example, if $a \notin S$ then there is always a hyperplane separating a from S . So we may want to have an oracle (called a *separation oracle* in this case) that improves upon the membership oracle inasmuch as if it answers $a \notin S$, then it also supplies us with the "reason", i.e. with the separating hyperplane.

The existence of separating hyperplanes for convex bodies is equivalent to the fact that each convex closed set is the intersection of all closed halfspaces containing it. So the convex set S could as well be described by specifying the set S^* of halfspaces containing it. This can again be done by a membership oracle for S^* , which is then called a *validity oracle* for S .

In the next section we formulate these oracles precisely, and also discuss what kind of "manufacturer's guarantees" are needed to make them algorithmically tractable (in the spirit of Section 1.4).

2.1. Algorithmic problems on convex sets.

Let $\mathcal{K}$ be a closed convex set in $\mathbb{R}^n$. Then there are two basic questions we can ask:

(2.1.1) MEMBERSHIP PROBLEM. Given a point $y \in \mathbb{Q}^n$, is $y \in \mathcal{K}$?

(2.1.2) VALIDITY PROBLEM. Given a halfspace $\{x : c^T x \le \gamma\} = H$ $(c \in \mathbb{Q}^n,\ \gamma \in \mathbb{Q})$, does it contain $\mathcal{K}$?

In both cases, if the answer is "No", we may want to have a "certificate" of this. One way to "certify" our answer is formulated in the following two problems.

(2.1.3) SEPARATION PROBLEM. Given a point $y \in \mathbb{Q}^n$, decide if $y \in \mathcal{K}$, and if not, find a hyperplane separating y from $\mathcal{K}$.

(2.1.4) VIOLATION PROBLEM. Given a halfspace H , decide if $\mathcal{K} \le H$, and if not, find a point in $H - \mathcal{K}$.

Of course, the first two questions can be asked for any set in $\mathbb{R}^n$; the reason for restricting our attention to closed convex sets is that in this case the MEMBERSHIP problem and the VALIDITY problem are "logically" equivalent: if we know the answer to either one of them for all inputs then we also know the answer to the other. The main result in this chapter will show that under appropriate additional hypotheses, these two problems are not only logically but also algorithmically equivalent. Also, the SEPARATION and VIOLATION problems will turn out algorithmically equivalent under reasonable conditions.

We show in some examples that these problems may be quite different for the same convex set.

(2.1.5) Let $\mathcal{K}$ be a polyhedron defined by a system of linear inequalities $a_i^T x \le b_i$ $(i = 1, \ldots, m)$. Then it is trivial to solve the separation problem and hence the membership problem for $\mathcal{K}$: just substitute the given point y in each of the given linear inequalities, to see if they are satisfied. On the other hand, the validity problem is equivalent to the solvability of a system of linear inequalities, which in turn is equivalent to linear programming. This can be solved by various algorithms, but far from easily.

(2.1.6) Let $\mathcal{K}$ be the convex hull of a given finite set of vectors. Then it is trivial to solve the violation problem and hence the validity problem for $\mathcal{K}$. On the other hand, the membership problem for $\mathcal{K}$ is again equivalent to a linear program.

(2.1.7) Let $f : \mathbb{R}^n \to \mathbb{R}$ be a convex function and $G_f = \{(x, t) : f(x) \le t\}$. Then the membership problem for G_f is trivial to solve, provided $f(x)$ can be computed for $x \in \mathbb{Q}^n$. To solve the separation problem for G_f , we need to be able to compute the subgradient of f . Concerning the validity problem for G_f , even to check whether halfspaces of the form $\{(x, t) : t \ge c\}$ contain G_f is equivalent to finding the minimum of f .

The above examples show that the difficulty of the problems formulated above depends very much on the way the convex body is given. Our approach will be to assume that $\mathcal{K}$ is given by an oracle which solves one of the problems above, and then study the polynomial solvability of the others.

Since we want to allow non–polyhedral convex sets, as well as polyhedra with irrational vertices, we have to formulate some weaker versions of these problems, where an error $\epsilon > 0$ is allowed. Let, for $\mathcal{K} \subseteq \mathbb{R}^n$, $S(\mathcal{K}, \epsilon)$ denote the ϵ–neighborhood of $\mathcal{K}$, i.e. $S(\mathcal{K}, \epsilon) = \{x \in \mathbb{R}^n : \inf_{y \in \mathcal{K}} ||x - y|| \le \epsilon\}$. We let $S(\mathcal{K}, -\epsilon) = \mathcal{K} - S(\mathbb{R}^n - \mathcal{K}, \epsilon)$. It helps to understand the definitions below if we read $y \in S(\mathcal{K}, -\epsilon)$ as "y is almost in $\mathcal{K}$ " and $y \in S(\mathcal{K}, \epsilon)$ as "y is deep in $\mathcal{K}$ ".

(2.1.8) WEAK MEMBERSHIP PROBLEM. Given a point $y \in \mathbb{Q}^n$ and a rational $\epsilon > 0$, conclude with one of the following:

(i) assert that $y \in S(\mathcal{K}, \epsilon)$;

(ii) assert that $y \notin S(\mathcal{K}, -\epsilon)$.

(2.1.9) WEAK SEPARATION PROBLEM. Given a vector $y \in \mathbb{Q}^n$, and a number $\epsilon \in \mathbb{Q}$, $\epsilon > 0$, conclude with one of the following:

(i) assert that $y \in S(\mathcal{K}, \epsilon)$;

(ii) find a vector $c \in \mathbb{Q}^n$ such that $||c||_\infty = 1$ and $c^T x < c^T y + \epsilon$ for every $x \in S(\mathcal{K}, -\epsilon)$.

(2.1.10) WEAK VALIDITY PROBLEM. Given a vector $c \in \mathbb{Q}^n$, a number $\gamma \in \mathbb{Q}$ and a number $\epsilon \in \mathbb{Q}$, $\epsilon > 0$, conclude with one of the following:

(i) assert that $c^T x \le \gamma + \epsilon$ for all $x \in S(\mathcal{K}, -\epsilon)$;

(ii) assert that $c^T x \ge \gamma - \epsilon$ for some $x \in S(\mathcal{K}, \epsilon)$.

(2.1.11) WEAK VIOLATION PROBLEM. Given a vector $c \in \mathbb{Q}^n$, a number $\gamma \in \mathbb{Q}$ and a number $\epsilon \in \mathbb{Q}$, $\epsilon > 0$, conclude with one of the following:
(i) assert that $c^T x \leq \gamma + \epsilon$ for all $x \in S(\mathcal{K}, -\epsilon)$;
(ii) find a vector $y \in S(\mathcal{K}, \epsilon)$ such that $c^T y > \gamma - \epsilon$.

Remark. Some of the ϵ's in the formulation of the "weak" problems above are redundant. The formulation above uses the weakest possible versions for the sake of uniformity and for later convenience. Of course, the two possible conclusions in either one of these problems are not mutually exclusive; there is a margin left for borderline cases when either output is legal.

To derive non-trivial relationships between these problems, we shall need some additional information on the convex sets. We shall restrict ourselves for most of this chapter to bounded and full-dimensional bodies. However, we want a guarantee for these properties. So if the convex set $\mathcal{K}$ is given by an oracle which solves one of the problems listed above, the black box will have to wear a guarantee:

(2.1.12)

The convex set described by this box is contained in $S(O, R)$.

and also one like

(2.1.13)

The convex set described by this box contains a ball with radius r.

Instead of this last guarantee, we could require other guarantees for the "non-flatness" of $\mathcal{K}$:

(2.1.14)

The convex set described by this box has width at least w

(the width of K is the minimum distance of two parallel hyperplanes with K between them) or

(2.1.15)

The convex set described by this box has volume at least v .

These three versions are, however, equivalent in the following sense. If (2.1.13) holds, then trivially (2.1.14) holds with $w = 2r$. Conversely, if (2.1.14) holds, then (2.1.13) holds with $r = w/(n+1)$. It is not hard to check that if (2.1.12) and (2.1.13) hold then we can compute from r, R and n in polynomial time a $v > 0$ for which (2.1.15) holds and vice versa, if (2.1.12) and (2.1.15) hold then we can compute from v, R and n in polynomial time an $r > 0$ for which (2.1.13) holds.

It will be convenient to assume that in these guarantees we have always $R \geq 1$ and $r, v, w \leq 1$. For our purposes, the interesting cases will be when r, v, w are very small and R is very large.

Sometimes we shall need a stronger guarantee instead of these: we need also the center of the inscribed ball.

(2.1.16)

The convex set described by this box contains the ball $S(a, r)$.

If an oracle describing a convex body has guarantees (2.1.12) and (2.1.13) we shall call it *well-guaranteed.* If it has guarantees (2.1.12) and (2.1.16), we shall call it *centered well-guaranteed.*

Of course, the boxes describing our convex sets have to wear guarantees that the sets are indeed convex and that different answers given by the oracle do not contradict each other. Also, the guarantees (2.1.12) – (2.1.16) have to be translated into explicit statements about the inputs and outputs of the oracle, depending on which of the above 8 problems is answered by the oracle. For example, a well–guaranteed validity oracle can have the following guarantees:

> For any two inputs $(c_1, \gamma_1), (c_2, \gamma_2)$
> if the answer is "yes" for both then it is
> also "yes" for the input $(c_1 + c_2;\ \gamma_1 + \gamma_2)$.
>
> For any input (c, γ) and any number $\lambda \in \mathbb{Q}_+$,
> if the answer is "yes" for (c, γ)
> then it is also "yes" for $(\lambda c, \lambda\gamma)$.
> For any input (c, γ) such that
> $\gamma > Rc$ the answer is "yes".
> For any input (c, γ) for which the answer
> is "yes", the answer to the input $(-c, -\gamma + w||c||)$
> is "no".

Accordingly, the *input size* of such an oracle is defined as $n + \langle R \rangle + \langle w \rangle$.

Let us mention two important problems which are closely related to the ones above:

(2.1.17) (LINEAR) OPTIMIZATION PROBLEM. Given a vector $c \in \mathbb{Q}^n$, find a vector $y \in \mathcal{K}$ such that $c^T y \geq c^T x$ for all $x \in \mathcal{K}$.

We obtain a more general problem if we replace the objective function by a more general concave function. Changing sign to obtain a more familiar form, let us assume that $\mathcal{K} \subseteq \mathbb{R}^n$ is a convex set and $f : \mathcal{K} \to \mathbb{R}$ is a convex function. Then we can formulate the following.

(2.1.18) CONSTRAINED CONVEX FUNCTION MINIMIZATION PROBLEM. Find a vector $y \in \mathcal{K}$ such that $f(x) \geq f(y)$ for all $x \in \mathcal{K}$.

Note that even the special case when $\mathcal{K} = \mathbb{R}^n$ is very important; this is called the *Unconstrained Convex Function Minimization Problem.*

Note, however, that the Constrained Convex Function Minimization Problem can be reduced to the *Linear* Optimization Problem easily: Just consider the set $\{(x,t) : x \in \mathcal{K}, t \geq f(x)\}$ and the linear objective function $(x,t) \mapsto t$.

The reader will have no difficulty in formulating the weak versions of these problems.

We also have to say some words about how the function f is given. We shall assume that it is given by an oracle whose input is a vector $x \in \mathcal{K}$ and a number $\epsilon \in \mathbb{Q}$, $\epsilon > 0$, and whose output is a number $r \in \mathbb{Q}$ such that $|f(x) - r| < \epsilon$. We shall call this a *weak evaluation oracle.*

2.2. The Ellipsoid Method.

The Ellipsoid Method is so well known that we shall only describe it briefly, omitting computational details. Instead, we shall spend more time on a lesser known version (namely on shallow cuts), because it is not so well known, but at the same time it is very important for many of the applications of the method.

The Ellipsoid Method is based on a method of Shor (1970) for convex function minimization using gradient projection with space dilatation. The method was described explicitly by Yudin and Nemirovskii (1976) and Shor (1977). Khachiyan (1979) adapted the method to show the polynomial time solvability of the linear programming problem. For an extensive survey of the topic, see Bland, Goldfarb and Todd (1981) and Schrader (1982) .

The basic ellipsoid method can be used to prove the following result.

(2.2.1) Lemma. *For a convex set $K \subseteq \mathbb{R}^n$ given by a well-guaranteed separation oracle, one can find a point in K in polynomial time.*

Proof. We construct a sequence of ellipsoids $E_0, E_1, \ldots, E_k$ such that $E_0 = S(O, R)$ (i.e. the ball occurring in guarantee (2.1.12)). If E_k is constructed, we inspect its center x_k using the separation oracle. If $x_k \in K$, we are done. If not, the separation oracle provides a vector $c_k \in \mathbb{Q}_n$ such that the halfspace $H_k = \{x : c_k^T x \leq c_k^T x_k\}$ contains K . Consider the half-ellipsoid $E_k \cap H_k$ and let E_{k+1} be an ellipsoid with minimum volume including $E_k \cap H_k$.

Then the following can be proved rather easily:

$$\text{(2.2.2)} \qquad \text{vol } (E_{k+1}) = \left(\frac{n^2}{n^2-1}\right)^{(n-1)/2} \frac{n}{n+1} \text{ vol } (E_k) < e^{-1/2n} \text{ vol } (E_k) .$$

From this observation it follows that the sequence of ellipsoids must terminate after a finite, and in fact polynomial, number of steps. Equation (2.2.2) implies that

$$\text{vol } (E_k) < e^{-k/2n} \text{ vol } (E_0) < e^{-k/2n}(2R)^n .$$

On the other hand, E_k contains K , and so by guarantee (2.5),

$$\text{vol } (E_k) \geq \text{ vol } (K) \geq v .$$

Hence

$$k \leq n^2 \log (2R) - n \log v .$$

This gives a polynomial upper bound on the number of ellipsoids in the sequence. It takes more work to see that the ellipsoids E_k can be computed in polynomial time. From a geometric point of view, this is still quite simple. We encode each ellipsoid E_k by a pair (A_k, x_k) where x_k is the center of E_k and A_k is a positive definite matrix such that

$$E_k = \{x \in \mathbb{R}^n : (x - x^k)^T A_k^{-1} (x - x_k) \leq 1\} .$$

Then it follows by a simple geometric argument that

$$(2.2.3)\qquad \begin{aligned} x_{k+1} &= x_k - \frac{1}{n+1} A_k c_k / \sqrt{c_k^T A_k c_k}\ , \\ A_{k+1} &= \frac{n^2}{n^2-1}\left(A_k - \frac{2}{n+1}\frac{A_k c_k c_k^T A_k}{c_k^T A_k c_k}\right) . \end{aligned}$$

The difficulties arise from the numerical calculations of the update formulas (2.2.3). They are twofold. First, the square root in (2.2.3) may result in irrational numbers, and so we cannot in general stay within the rational field. Hence one has to round, and it takes a somewhat tedious computation to show that it suffices to round to a polynomial number of digits. Second, one has to make sure that the input size of the numbers which arise remains bounded by some polynomial of the input size. This can also be done by carefully estimating the norms of the matrices A_k and A_k^{-1} . We suppress these considerations here and refer instead e.g. to Grötschel, Lovász and Schrijver (1981).

Since rounding is involved anyway, it turns out that it suffices to have a weak separation oracle. Of course, in that case we can find only a vector $y \in S(K, \epsilon)$ with the method, for any prescribed $\epsilon > 0$.

To attack the violation problem for K , we may replace K by the convex set $\overline{K-H}$, for which a separation oracle is trivially constructed from a separation oracle for K . However, this separation oracle is not well-guaranteed in general; no lower bound on the volume (or width) or $\overline{K-H}$ can be derived immediately. So we have to add to the Ellipsoid Method as described above a new *stopping rule*: if vol $E_k < \epsilon'$ for an appropriate prescribed ϵ' then we also stop.

Now the ellipsoid method may have two outcomes: either it stops with a point in K , or it stops with a certificate for vol $(K-H) < \epsilon'$. In this second case, we can be sure that $S(K, -\epsilon) \subseteq H$ if we have chosen $\epsilon' = (\frac{\epsilon}{2R})^n v$. So the method yields a solution to the weak violation problem. Since again a weak separation oracle was sufficient, we have sketched the proof of the following result, which is fundamental for the sequel.

(2.2.4) Lemma. *For a convex set given by a well-guaranteed weak separation oracle, the weak violation problem can be solved in polynomial time.*

Let us remark that if we can solve the weak violation problem, then, using binary search, a polynomial algorithm to solve the weak optimization problem is easily obtained. Hence

(2.2.5) Corollary. *For a convex set given by a well-guaranteed weak separation oracle, the weak optimization problem can be solved in polynomial time.*

Our next goal is to prove a converse of Lemma (2.2.4). This will be relatively easy by a polarity argument, but first we have to find a point "deep in K ", which can then serve as the center of the polarity. This is the purpose of the next lemma.

(2.2.6) Lemma. *For a convex set $K \subseteq \mathbb{R}^n$ given by a well–guaranteed weak violation oracle, we can determine in polynomial time a vector $a \in \mathbb{Q}^n$ and a number $r' \in \mathbb{Q}$, $r' > 0$ such that $S(a, r') \subseteq K$.*

Proof. (Sketch.) We choose a very small $\epsilon > 0$ and then select affinely independent points $x_0, \dots, x_n \in S(K, \epsilon)$ as follows. First, by calling the weak violation oracle with any $c \neq 0, \gamma \leq -R\|c\|$ and the given ϵ , we obtain a vector $x_0 \in S(K, \epsilon)$. If $x_0, \dots, x_i$ are already selected $(i < n)$, then let $c \neq 0$ be any vector such that $c^T x_0 = \dots = c^T x_i$, and call the weak violation oracle with $(c,\ c^T x_i + \frac{w}{3}\|c\|,\ \epsilon)$ and $(-c,\ -c^T x_i + \frac{w}{3}\|c\|,\ \epsilon)$, where w is the lower bound on the width of K guaranteed by (2.3). Then in at least one case the oracle supplies a point $x_{i+1} \in S(K, \epsilon)$.

Now if $x_0, \dots, x_n$ are found, then we can let $S(a, r'')$ be the ball inscribed in the simplex $\text{conv}\{x_0, \dots, x_n\}$. Then with $r' = r'' - \epsilon$, we will have that $S(a, r') \subseteq K$. □

We can prove now the converse of Lemma (2.2.4).

(2.2.7) Lemma. *For a convex set given by a well–guaranteed weak violation oracle, the weak separation problem can be solved in polynomial time.*

Proof. Let $K \subseteq \mathbb{R}^n$ be our convex set. Using Lemma (2.2.6), we can find in polynomial time a vector $a \in \mathbb{Q}^n$ and a number $r' \in \mathbb{Q}$, $r' > 0$ such that $S(a, r') \subseteq K$. For notational convenience, we assume that $a = 0$ and $r' = r$. Let $K^* = \{x \in \mathbb{R}^n : x^T y \leq 1 \text{ for all } y \in K\}$ be the *polar* of K .

Let us note first that

$$S\left(0, \frac{1}{R}\right) \subseteq K^* \subseteq S\left(0, \frac{1}{r}\right) ,$$

and hence guarantees that (2.1.12) and (2.1.13) (even (2.1.16)) can easily be supplied for K^* .

Furthermore, the separation problem for K^* is easily solved in polynomial time. Let $y \in \mathbb{Q}^n$ and $\epsilon \in \mathbb{Q}, \epsilon > 0$ be given. Call the validity oracle for K with the input $(y, 1, \epsilon')$ where $\epsilon' = \frac{\epsilon r}{2(1+\|y\|)^2}$. If it finds that $y^T x \leq 1 + \epsilon'$ for all $x \in S(K, -\epsilon')$, then $y^T x \leq 1 + \epsilon'\|y\|$ for all $x \in K$ and hence $\frac{1}{1+\epsilon'+\epsilon'\|y\|} \cdot y \in K^*$ by the definition of K^* . Hence $y \in S(K^*, \epsilon)$.

On the other hand, if the weak violation oracle for K gives a vector $u \in S(K, \epsilon')$ such that $y^T u > 1 - \epsilon'$ then let $c = u/\|u\|$. We claim that c is a valid output for (ii) in the weak separation problem for K^* . Let $x \in S(K^*, -\epsilon)$. Then in particular $x \in K^*$. Since $u \in S(K, \epsilon')$, there is a vector $u_0 \in K$ such that $\|u - u_0\| \leq \epsilon'$. Hence

$$u^T x \leq \|u - u_0\| \cdot \|x\| + |u_0^T x| \leq 1 + \epsilon' \cdot \frac{1}{r} ,$$

and so

$$c^T x \leq \left(1 + \epsilon' \frac{1}{r}\right) \cdot \frac{1}{\|u\|} .$$

Thus

$$c^T x - c^T y \le \left(1 + \epsilon' \cdot \frac{1}{r}\right) \cdot \frac{1}{||u||} + (1 - \epsilon')\frac{1}{||u||} = \epsilon' \cdot \frac{1}{||u||}\left(\frac{1}{r} + 1\right) \le \epsilon .$$

This completes the proof of the fact that the weak separation problem for K^* is solvable in polynomial time.

So by Lemma (2.2.4), the violation problem for K^* can be solved in polynomial time. But interchanging the role of K and K^*, the preceding argument then shows that the separation problem for K^* can be solved in polynomial time. □

Next we turn to the study of the membership oracle. It is clear that the membership oracle is weaker than the separation oracle; it does not even suffice to find a point in the set $S(K, \epsilon)$ in polynomial time. This can be formulated precisely as follows.

(2.2.8) Theorem. *Every algorithm which finds, for every convex set K given by a well-guaranteed (weak or strong) membership oracle and for every $\epsilon > 0$, a vector in $S(K, \epsilon)$ in polynomial time, takes an exponential number of steps in the worst case.*

Proof. We construct an oracle which even promises to describe one of the m^n little cubes (not telling, of course, which one) obtained by splitting the unit cube into cubes of size $1/m$, where m is any natural number. From this promise, guarantees (2.1.12) and (2.1.13) are easily derived. Let us design the inside of the box, however, so that for any membership query it should answer "no" as long as there is at least one little cube no point of which was queried. So its answer to the first $m^n - 1$ questions in particular will be no, and it takes at least $m^n - 1$ questions before the algorithm can find a point which is certainly in K. □

Remark. Note that m^n is exponentially large in the input size $\langle m \rangle + n$ even if m or n is fixed!

It is quite surprising that if we strengthen the guarantee just a little, and require (2.1.16) instead of (2.1.13) (i.e. if we assume that the center of the little ball inside K is explicitly given by the manufacturer of the black box for the weak membership oracle), then the membership problem is in fact equivalent to the weak separation (or violation) problems. This fact was first proved by Yudin and Nemirowskii (1976). We shall sketch the proof, which involves a little-known version of the Ellisoid Method, the so-called Shallow Cut Ellipsoid Method. However, we shall have to skip computational details.

Since the membership problem is a relaxation of the separation problem, it is a natural idea to use a (weak) membership oracle to obtain a (weak) separation. However, no direct way to achieve this is known. What we can do is to obtain a "very weak separation" from the membership oracle, and then strengthen the Ellipsoid Method so that even this "very weak separation" should be enough for it. More precisely, we prove the following result.

(2.2.9) Lemma. *For a convex set $K \subseteq \mathbb{R}^n$ given by a centered, well-guaranteed membership oracle, for any vector $y \in \mathbb{Q}^n$ and any number $\epsilon \in \mathbb{Q}, \epsilon > 0$, one can reach one of the following two conclusions in polynomial time:*

(i) *assert that $y \in S(K, \epsilon)$;*

(ii) *find a vector $c \in \mathbb{Q}^n$ such that $||c|| = 1$, and for every $x \in K$,*

$$c^T x \leq c^T y + \epsilon + \frac{1}{n+1}||x - y|| .$$

Proof. (Sketch.) It may be easier to see the geometric content of (ii) if we remark that a vector c such that $||c|| = 1$ and

$$c^T x \leq c^T y + \frac{1}{n+1}||x - y|| \quad (x \in K)$$

is just a unit vector such that the rotational cone with vertex in y, axis parallel to c and semi-angle $\operatorname{arctg}(n+1)$ (which is quite close to $\pi/2$) is disjoint from K. With the ϵ in the statement, it suffices to find a congruent cone with vertex closer to y than ϵ and disjoint from K; and this is what we shall achieve.

We assume, to make the main idea behind the algorithm more clear, that we have a *strong* membership oracle; with a weak oracle, it would take some additional work but no essentially new idea to carry out the calculations.

First, we call the membership oracle to test if $y \in K$; if the answer is "yes", we are done, so suppose that $y \notin K$.

By binary search along the segment connecting the "center" point a in guarantee (2.1.16) and y, we can find two vectors p and q and a small but positive (and polynomial computable) $\delta > 0$ such that $||p - q|| < \frac{\epsilon}{2}$ and $S(p, \delta) \subseteq K$ but $q \notin K$. For notational convenience, suppose that $p = 0$ and $q = y$.

Let $u_1, \ldots, u_{n-1} \in \mathbb{R}^n$ be vectors orthogonal to y and to each other, such that $||u_i|| = ||y||$. Consider the $2n-2$ points $v_i = \alpha y + \beta u_i$ and $w_i = \alpha y - \beta u_i$, where $\alpha = \frac{n^2(n+2)}{n^3+2n+1}$, $\beta = \frac{n\sqrt{n+2}}{n^3+2n^2+1}$. (So 0, y and any v_i from a narrow rectangular triangle with the rectangle at point v_i, and similarly for w_i.) We call the membership oracle for each of the points v_i and w_i, and distinguish two cases.

Case 1. For all i, $v_i \in K$ and $w_i \in K$. Then we claim that no vector in the cone

$$C = y + \left\{ \sum_{i=1}^{n-1} \lambda_i (y - v_i) + \sum_{i=1}^{n-1} \mu_i (y - w_i) : \lambda_i, \mu_i \geq 0 \right\}$$

belongs to K. Assume that $u = y + \sum_{i=1}^{n-1} \lambda_i(y - v_i) + \sum_{i=1}^{n-1} \mu_i(y - w_i) \in K$ for some $\lambda_i, \mu_i \geq 0$. Clearly not each λ_i and μ_i is 0. Let $\sum_{i=1}^{n-1} \lambda_i + \sum_{i=1}^{n-1} \mu_i = t$ and consider the vector

$$z = \frac{1}{t} \left(\sum_{i=1}^{n-1} \lambda_i v_i + \sum_{i=1}^{n-1} \mu_i w_i \right) .$$

By the convexity of K, $z \in K$ and hence also $\frac{t}{1+t}z + \frac{1}{1+t}u \in K$. But $\frac{t}{1+t}z + \frac{1}{1+t}u = y$, which is a contradiction.

Also note that C contains the rotational cone with center y , axis y and half-angle $\text{arctg}(n+1)$. So by the remark after the statement of the lemma, we are done.

Case 2. Suppose that e.g. $v_i \notin K$. Then we replace y by v_i and repeat the procedure.

It remains to show that after a polynomial number of iterations in Case 2, Case 1 must occur; and also that the vertices of the cones do not "drift away" from y . Both of these facts follow from the observation that

$$||v_i|| = \sqrt{\frac{n^2(n+2)}{n^2(n+2)+1}}||y|| \le e^{-1/2(1/n^2(n+2)+1)}||y|| .$$

So all cone vertices remain no further away from y than $2||y|| \le \epsilon$; and the number of iterations is at most

$$2(n^3+2n^2+1)\,|\log \epsilon|\,/\,|\log \delta| \ ,$$

which is polynomial in the input size. □

Remark. We could replace the coefficient $\frac{1}{n+1}$ in (ii) by any number $\frac{1}{p(n)}$, where p is a polynomial. What will be important for us, however, is that it is smaller than $\frac{1}{n}$.

(2.2.10) Lemma. *For a convex set given by a centered, well-guaranteed weak membership oracle, we can solve the weak violation problem in polynomial time.*

Proof. Let a halfspace H and a number $\epsilon > 0$ be given. We proceed as in the basic version of the Ellipsoid Method: we construct a sequence $E_0, E_1, \ldots$ of ellipsoids such that $K - H \subseteq E_k$ and $\text{vol}(E_k)$ decreases. The sequence terminates if the center x_k of E_k is in $K - H$. Again, we can take $E_0 = S(O,R)$.

Suppose that E_k is constructed but its center $x_k \notin K - H$. If $x_k \in H$ then we include $E_k - H$ in an ellipsoid E_{k+1} just like in the basic ellipsoid method and we have $\text{vol}\,(E_{k+1}) < e^{-\frac{1}{2(n+1)}}\ \text{vol}\ (E_k)$. So suppose that $x_k \notin H$. Then $x_k \notin K$.

Let Q be an affine transformation mapping E_k onto the unit ball $S(0,1)$. Then $0 \notin QK$. Furthermore, it is easy to design a centered well-guaranteed weak membership oracle for QK . So we can use Lemma (2.2.9) to find a "very weak separation" for 0 , i.e. a vector $c \in \mathbb{R}^n$ such that $||c|| = 1$ and $c^Tx < \epsilon + \frac{1}{n+1}||x||$ for all $x \in K$.

Consider now the set

$$B' = \left\{ x \in \mathbb{R}^n : ||x|| \le 1,\ c^Tx \le \frac{2}{2n+1} \right\} . \tag{2.2.11}$$

By the choice of c, B' contains K. Now B' is more than a half ball; still, it can be included in an ellipsoid E'_{k+1} such that vol E'_{k+1} $<$ vol $S(0,1)$. In fact,

$$\text{vol }(E'_{k+1}) < e^{-3/2(n+1)(2n+1)^2} \text{ vol } S(0,1)$$

by lengthy but elementary computations. Hence if we take $E_{k+1} = Q^{-1}E'_{k+1}$, we have

$$\text{vol }(E_{k+1}) < e^{-2/2(n+1)(2n+1)^2} \text{vol }(E_k) .$$

This shrinking factor is closer to 1 than in the basic ellipsoid method, but it still suffices to show that the procedure must terminate in a polynomial number of steps.

(The crucial fact that B' can be included in an ellipsoid whose volume is smaller than the volume of $S(0,1)$ depends on the fact that the number $\frac{2}{2n+1}$ occurring in (2.2.11) is smaller than $\frac{1}{n}$. If we replace this number by any number $\geq \frac{1}{n}$, the ellipsoid with smallest volume containing B' would be $S(0,1)$ itself.) □

Using a polarity argument similar to the one in the proof of Lemma (2.2.5), we can derive that for a convex set given by a centered, well–guaranteed weak validity oracle, we can solve the weak separation problem in polynomial time. In this result, however, we do not need the hypothesis that we know the "center".

(2.2.12) Lemma. *For a convex set given by a well–guaranteed weak validity oracle, we can solve the weak separation problem in polynomial time.*

Proof. Let $K \subseteq \mathbb{R}^n$ be the given set. Embed $\mathbb{R}^n$ into $\mathbb{R}^{n+1}$ by appending an $(n+1)^{\text{st}}$ coordinate. Let $a = \binom{0}{1} \in \mathbb{R}^{n+1}$ and let $\hat{K} = \text{conv }(K \cup S(a, 1/2))$. Then it is trivial to design a weak validity oracle for $\hat{K}$, and also to supply it with proper guarantees. A ball contained in $\hat{K}$ is also given, so by the remark above, we can solve the weak separation problem for $\hat{K}$.

Now, let $y \in \mathbb{R}^n$ and $\epsilon > 0$ be given. Let $\hat{y} = \binom{y}{t}$ where $t = \epsilon/5R$. Call the weak separation algorithm for $\hat{K}$ with very small error δ. If this concludes that $\hat{y} \in S(\hat{K}, \delta)$ then there follows an easy geometric argument that $y \in S(K, \epsilon)$. On the other hand, if it finds a separating hyperplane, then this intersects $\mathbb{R}^n$ in a hyperplane of $\mathbb{R}^n$ separating y from K. (Again we have suppressed some technical details, arising mainly from the fact that the separating hyperplane in $\mathbb{R}^{n+1}$ may intersect $\mathbb{R}^n$ in a rather small angle. But this can be handled.) □

Finally, consider optimization problems.

(2.2.13) Lemma. *For a convex set given by a weak violation oracle, the weak linear optimization problem can be solved in polynomial time.*

Proof. Let $c \in \mathbb{Q}^n$ and $\epsilon \in \mathbb{Q}$, $\epsilon > 0$, and say, $||c|| = 1$. By binary search, we can find a number $\gamma \in \mathbb{Q}$ such that $c^T x \leq \gamma$ is found "almost valid" by the oracle but $c^T x \leq \gamma - \frac{\epsilon}{3}$ is found "almost violated", with error $\frac{\epsilon}{3}$. This means that the oracle has asserted that $c^T x \leq \gamma + \frac{\epsilon}{3}$ is valid for all $x \in S(K, -\frac{\epsilon}{4})$ but that it also supplies a vector $y \in S(K, \frac{\epsilon}{3})$ such that $c^T y \geq \gamma - \frac{2\epsilon}{3}$. So it follows that $y \in S(K, \epsilon)$ and that $c^T y \geq \gamma - \frac{2\epsilon}{3} \geq c^T x - \epsilon$ for all $x \in S(K, -\epsilon)$, and so y is a solution to the weak linear optimization problem. □

As a corollary, we find that the weak linear optimization problem can be solved in polynomial time if we have a well–guaranteed separation, violation, validity or centered membership oracle. Since, conversely, an oracle for the weak linear optimization problem trivially answers the weak violation problem, we obtain the following result.

(2.2.14) Theorem. *For a convex set K , the following oracles, when well–guaranteed, are polynomially equivalent.*

(a) *centered weak membership;*
(b) *weak validity;*
(c) *weak separation;*
(d) *weak violation;*
(e) *weak linear optimization.*

We have also seen that if we drop "centered" from (a) then we obtain a strictly weaker oracle.

We conclude this section with the following important application of the preceding results.

(2.2.15) Theorem. *Let $K \subseteq \mathbb{R}^n$ be a convex set given by any well–guaranteed oracle of the list in Theorem (2.2.14). Let $f : K \to \mathbb{R}$ be a convex function given by a weak evaluation oracle. Then the weak constrained function minimization problem for K and f can be solved in polynomial time.*

Proof. By our general hypothesis on oracles, we can compute in polynomial time an upper bound μ on max $\{|f(x)| : x \in K\}$. Consider the set

$$\hat{K} = \{(x, \epsilon) : x \in K,\ f(x) \leq t \leq \mu + 1\} .$$

Clearly, $\hat{K}$ is a convex set. It is trivial to design a membership oracle for $\hat{K}$. It is also easy to supply this with guarantees (2.1.12) and (2.1.16). So by Theorem (2.2.14), the weak linear optimization problem for K is polynomially solvable. But then in particular we can find in polynomial time for every $\epsilon > 0$ a vector (y, s) that "almost minimizes" the linear objective function $(x, t) \mapsto t$. Then y is a solution of the weak constrained convex function minimization problem. □

2.3. Rational polyhedra.

Let $P \subseteq \mathbb{R}^n$ be a polyhedron and ϕ and ν positive integers. We say that P has *facet complexity* at most ϕ if P can be described as the solution set of a system of linear inequalities each of which has input size $\leq \phi$. We say that P has *vertex complexity* at most ν if P can be written as $P = \text{conv}\,(V) + \text{cone}\,(E)$, where $V, E \subseteq \mathbb{Q}^n$ are finite and each vector in $V \cup E$ has input size $\leq \nu$. By an elementary computation one can see that ϕ and ν can be estimated by polynomials of each other. More exactly, if P has facet complexity $\leq \phi$ then it has vertex complexity $\leq 4\phi^3$ and conversely, if it has vertex complexity $\leq \nu$ then it has facet complexity $\leq 4\nu^3$.

So the facet and vertex complexities are in a sense equivalent, and we shall use whichever is more convenient. It is also easy to see that $n \leq \nu, \phi$.

If we know an upper bound on the facet (or vertex) complexity of P , then various other properties can be derived. The following two are the most important.

(2.3.1) Lemma. *If P is a polytope (i.e. a bounded polyhedron) with vertex complexity $\leq \nu$, then $P \subseteq S(0, 2^\nu)$.*

(2.3.2) Lemma. *If P is a full-dimensional polyhedron with facet complexity ϕ , then it contains a ball with radius $2^{-7\phi^3}$.*

If a polyhedron P is given by some oracle, then a *rationality guarantee* for P is one of the following form:

The polyhedron described by this box has facet complexity $\leq \phi$

(we could equivalently guarantee that the vertex complexity is small). We consider here ϕ in *unary encoding.* So Lemmas (2.3.1) and (2.3.2) imply that for a full-dimensional bounded polyhedron with rationality guarantee, we can give guarantees of the form (2.1.12) and (2.1.13).

Our main result on rational polyhedra shows that a rationality guarantee enables us to strengthen the main result of the previous section (Theorem (2.2.14)) in several directions: for the last three problems, the weak and strong versions are equivalent, and the boundedness and full-dimensionality hypotheses can be dropped.

(2.3.3) Theorem. *For a polyhedron P the following oracles, with rationality guarantee, are polynomially equivalent:*

(a) *strong separation,*
(b) *strong violation,*
(c) *strong linear optimization.*

If P is full-dimensional, then the following oracles (with rationality guarantee) are also polynomially equivalent with (a):

(d) *weak separation,*
(e) *weak violation,*

(f) *weak linear optimization,*
(g) *centered weak membership,*
(h) *centered strong membership.*
If P *is bounded then*
(i) *strong validity,*
if P *is both bounded and full–dimensional then also*
(j) *weak validity*
(with rationality guarantee) is equivalent to (a).

We shall not give the proof of this theorem in detail; in the outline which follows, we shall expose the points where the difficulty lies.

First, it follows from Theorem (2.2.14) that for bounded, full–dimensional polyhedra the weak oracles (d), (e), (f), (g) and (j) are polynomially equivalent. It is not difficult to extend these results in the first three cases to the unbounded case. Next one shows that in the full-dimensional case, the strong versions of these problems are also equivalent. This can be done by finding appropriate approximations to the solutions and then by careful rounding. At this point, one can make use of the simultaneous diophantine approximation algorithm in the preceding chapter. The main tool is the following reformulation of Theorem (1.3.6):

(2.3.4) Lemma. *Let* P *be a polyhedron with facet complexity* $\leq \rho$, *and* $y \in \mathbb{Q}^n$, *such that* $y \in S(P, 2^{-4n\rho})$; *then we can compute in polynomial time a vector* $\overline{y} \in P$ *such that* $||y - \overline{y}|| \leq 2^{-2\rho}$. □

By a similar method, if we have a linear inequality $a^T x \leq \alpha$ that is "almost valid" for P , then by simultaneous diophantine approximation, we can find a valid inequality for P .

It is interesting to note that the rounding procedure in both cases is independent of P (it only depends on the facet or vertex complexity of P) . One may also check that such a rounding cannot be achieved by rounding each entry individually.

Using the rounding procedures described above, it is not hard to show that weak and strong optimization as well as weak and strong violation are equivalent for full–dimensional polyhedra. The case of weak and strong separation is slightly more involved, but one can use the Ellipsoid Method to find a little ball inside P , and then apply polarity just as in the proof of Lemma (2.2.7).

The most difficult part is to deal with non–full–dimensional polyhedra. The key step is to determine their affine hull; then the problem is reduced to the full–dimensional case.

If the polyhedron P is given by a strong violation or strong optimization oracle (with rationality guarantee), then the affine hull of P can be found in polynomial time using the same sort of procedure as in the proof of Lemma (2.2.6) (Edmonds, Lovász and Pulleyblank (1982)).

Suppose now that P is given by a strong separation oracle. Assuming that the complexity of the output is uniformly bounded, Karp and Papadimitriou (1981) proved that the strong optimization problem is solvable then for P . A similar result was obtained by Padberg and Rao (1984) under the related hypothesis that the oracle gives facets as separation. To obtain this result without any assumption about the oracle, we use simultaneous diophantine approximation again.

So let P be given by a strong separation oracle. We want to find the affine hull of P. It is easy to see that one may assume that P is bounded.

We start with the basic ellipsoid method and construct a sequence $E_1, E_2, \ldots$ of ellipsoids such that $P \subseteq E_k$ and vol $E_k \to 0$. If we find that the center x_k of E_k belongs to P, then we check also the points $x_k + \epsilon e_i$, where $\epsilon > 0$ is very small and $e_1, \ldots, e_n$ are the basis vectors of $\mathbb{R}^n$. If we find that all these n points are in P then P is full-dimensional and so its affine hull is $\mathbb{R}^n$. So we may assume that we always find a vector $x_k + \epsilon e_i \notin P$. Then from the separation oracle we obtain a hyperplane separating $x_k + \epsilon e_i$ from P, and we can use this as a "slightly shallow" cut to obtain E_{k+1}.

In a polynomial number of steps we have that vol $(E_k) < 2^{-7\nu n^2}$. Then the width of E_k is less than $2^{-6\nu n}$, i.e. there is a hyperplane $a^T x = \alpha$ such that both inequalities $a^T x \leq \alpha + 2^{-6\nu n}$ and $a^T x \geq \alpha - 2^{-6\nu n}$ are valid for E_k and, consequently, also valid for P. But then rounding a and α as sketched above using simultaneous diophantine approximation, we obtain a hyperplane $c^T x = \nu$ which contains P. From here we can go on by induction on the dimension of the space and find the affine hull of P in polynomial time.

Several other algorithmic questions concerning polyhedra can be settled using these methods. Let us list some:

(a) Find a vertex (if any).

(b) Given a vector $y \in P$, find vertices $v_0, \ldots, v_n$ and numbers $\alpha_a, \ldots, \alpha_n$ such that $\alpha_i \geq 0$, $\sum \alpha_i = 1$ and $\alpha_0 v_0 + \ldots + \alpha_n v_n = y$.

(c) Given a valid inequality $c^T x \leq \nu$, find facets $a_i^T x \leq \alpha_i$ $(i = 1, \ldots, n)$ and numbers π_i $(i = 1, \ldots, n)$ such that $\pi_i \geq 0$, $\sum \pi_i a_i = c$, $\sum \pi_i \alpha_i \leq \nu$.

(d) Given a point $y \notin P$, find a facet $a^T x \leq \alpha$ such that $a^T y > \alpha$.

(2.3.5) Theorem. *If P is a polyhedron given by a strong separation (violation, optimization) oracle with rationality guarantee, then problems* (a)–(d) *above can be solved in polynomial time.* □

We use the methods of this section to derive one, seemingly technical, strengthening of the previous results. This is based on results of Tardos (1984) and Frank and Tardos (1985).

(2.3.6) Theorem. *Let $P \subseteq \mathbb{R}^n$ be a polyhedron given by a strong separation oracle with rationality guarantee, and let $c \in \mathbb{Q}^n$. Then a vertex of P maximizing the linear objective function $c^T x$ over P can be found in time which is polynomial in the input size of the oracle and strongly polynomial in c (i.e. the number of arithmetic and other operations does not depend on c, only on the input size of the numbers on which the operations are performed).*

Before proving this result, some comments on its significance are in order. First, one obtains as a special case:

(2.3.7) Corollary. *The linear program*

$$\text{(2.3.8)} \qquad \begin{array}{l} \textit{maximize } \ c^T x \\ \textit{subject to } \ Ax \le b \end{array}$$

can be solved by an algorithm which is polynomial in $< A > + < b >$ *and strongly polynomial in* c .

By applying a similar argument also to the dual program, one can prove even more:

(2.3.9) Theorem. *The linear program* (2.3.8) *can be solved by an algorithm which is polynomial in* $< A >$ *and strongly polynomial in* b, c .

A more direct proof of this result, avoiding simultaneous diophantine approximation, has also been found by Tardos (1985) .

In many cases, especially in those arising in combinatorial optimization, the matrix A has very simple entries, often 0's and ± 1's. Hence in these cases the strong polynomiality of the algorithm follows. In particular, one finds as a corollary that the Minimum Cost Flow Problem can be solved in strongly polynomial time (see Chapter 3).

Let us sketch a proof of Theorem (2.3.6). We may assume that $||c||_\infty = 1$. For brevity, assume that P is bounded, and let P have vertex complexity $\le \nu$. By Theorem (1.3.7) we can find a vector $\tilde{c} \in \mathbb{Q}^n$ such that $\langle \tilde{c} \rangle \le 12\nu n^4$ and every strict linear inequality with input size $\le 2\nu$ satisfied by c is also satisfied by $\tilde{c}$. In particular, if $c^T v_1 > c^T v_2$ for two vertices of P then $\tilde{c}^T v_1 > \tilde{c}^T v_2$ and vice versa. So the same vertices of P maximize c and $\tilde{c}$. Hence it suffices to maximize $\tilde{c}^T x$ over P . But $\langle \tilde{c} \rangle$ is bounded independently of $\langle c \rangle$. □

2.4. Some other algorithmic problems on convex sets.

Let K be a convex set given by a well-guaranteed oracle for the weak separation, violation, validity or optimization problem, or by a centered weak membership oracle. Since all these are polynomially equivalent, we shall not distinguish between them, and will call K given by any of them a *convex body.*

It was proved by Löwner (cf. Danzer, Grünbaum, and Klee (1963)) and John (1948) that for every convex body $K \subseteq \mathbb{R}^n$ there exist a pair (E', E) of ellipsoids such that $E' \subseteq K \subseteq E$, E' and E are concentric, and E' arises from E by shrinking by a factor $1/n$. In fact, if we choose E to be the smallest volume ellipsoid including K , or E' the largest volume ellipsoid contained in K , then such a pair arises. We call such a pair a *Löwner–John pair* for K .

The main result in this section is that an approximation of a Löwner–John pair can be computed for every convex body in polynomial time. This result can then be applied to finding approximations for other important geometric parameters of K.

Let us call a pair (E', E) of ellipsoids a *weak Löwner–John pair* for K if $E' \subseteq K \subseteq E$, E' and E are concentric and E' arises from E by shrinking by a factor $1/(n+1)\sqrt{n}$.

(2.4.1) Theorem. *Let $K \subseteq \mathbb{R}^n$ be a convex body. Then a weak Löwner–John pair for K can be computed in polynomial time.*

Proof. Let us assume that K is given by a weak separation oracle. We shall in fact assume that the oracle is strong; this is not justified but the proof will be clearer. We use the shallow cut ellipsoid method. We construct a sequence of ellipsoids $E_0, E_1, \ldots$ such that $K \subseteq E_k$ for all k and

$$\text{vol}\,(E_{k+1}) < e^{-3/2(n+1)(2n+1)^2}\,\text{vol}\,(E_k)\,. \tag{2.4.2}$$

This sequence must terminate in a polynomial number of steps. Suppose we have E_k. We show that either E_k determines a weak Löwner–John pair for K or we can construct E_{k+1}.

First, check if the center x_k of E_k belongs to K. If not, we proceed by the basic ellipsoid method. Suppose that $x_k \in K$. Next, determine the endpoints of the axes of E_k; let these points be $x_k \pm a_i$ $(i = 1, \ldots, n)$. Check if $x_k \pm \frac{1}{n+1} a_i \in K$, for $i = 1, \ldots, n$. If all these $2n$ points are in K, then so is their convex hull P. But P contains the ellipsoid E'_k with center x_k obtained from E_k by shrinking by a factor $1/(n+1)\sqrt{n}$. So (E'_k, E_k) is a weak Löwner–John pair.

Suppose now that e.g. $x_k + \frac{1}{n+1} a_1 \notin K$. Then the separation oracle gives a halfspace H such that $K \subseteq H$ but $x_k + \frac{1}{n+1} a_1 \notin H$. Such a H is a "shallow cut", i.e. if we let E_{k+1} be an ellipsoid with smallest volume including $E_k \cap H$, then (2.4.2) holds, and we are done. □

Remarks. 1. If we apply an affine transformation which maps the weak Löwner–John pair onto balls, then the image of K is reasonably "round" in the sense that its circumscribed ball is at most $(n+1)\cdot\sqrt{n}$ times larger than its inscribed ball. This is to what the chapter title refers.

2. To see why (2.4.2) holds, one may apply an affine transformation mapping E_k on $S(0,1)$. Then (2.4.2) is just the same formula as in the proof of Lemma (2.2.10).

3. For special convex bodies K, one may improve the factor $1/(n+1)\sqrt{n}$. If K is centrally symmetric, then one obtains $1/n$. If K is a polytope given either as the solution set of a system of linear equations or as the convex hull of a set of vectors, then $1/(n+1)$ can be achieved. Finally, if K is a centrally symmetric polytope and is given as above, then $1/\sqrt{n+1}$ can be achieved.

Let us turn to other geometric problems. Perhaps the most important is the *volume*. Theorem (2.4.1) implies immediately:

(2.4.3) Corollary. *For every convex body* K , *one can compute in polynomial time a number* $\mu(K)$ *such that*

$$\mu(K) \leq \text{vol}(K) \leq n^{n/2}(n+1)^n \mu(K) .$$

Proof. Let $\mu(K) = \text{vol}\,(E')$ for any Löwner–John pair (E', E) for K. □

The coefficient $n^{n/2}(n+1)^n$ seems to be outrageously bad. But the following result of G. Elekes (1982) shows that every polynomial time algorithm must leave a very bad margin for error.

(2.4.4) Theorem. *There is no polynomial time algorithm which would compute a number* $\overline{\mu}(K)$ *for each convex body* K *such that*

$$\overline{\mu}\ (K) \leq \text{vol}(K) \leq 1.999^n \overline{\mu}(K) .$$

Proof. Let $S(0,1)$ be the unit ball in $\mathbb{R}^n$, v_n its volume, and let $v(n,p)$ denote the maximum volume of a polytope representable as the convex hull of p points in $S(0,1)$.

Suppose that we have an algorithm which computes an upper bound on the volume in polynomial time. We shall restrict ourselves to convex bodies contained in $S(0,1)$ and containing $S(0,1/2)$, and hence we may assume that the bodies in consideration are given by a membership oracle.

Let us apply our algorithm to $S(0,1)$, and assume that it finds the value $\overline{\mu}(S(0,1))$. Let $v_1, \ldots, v_p$ be those points in $S(0,1)$ whose membership has been tested; since the algorithm is polynomial, p is bounded by a polynomial in n (the other input parameters being constant : $R = 1$ and $r = \frac{1}{2}$).

Now if we run our algorithm with $K = \text{conv}\ \{v_1, \ldots, v_p\}$, we obtain the same answers and so the result will be the same number:

$$\overline{\mu}(K) = \overline{\mu}(s(0,1)) .$$

Let $\text{vol}(S(0,1)) = v_n$, then we have that

$$\overline{\mu}(K) \leq \text{vol}(K) \leq 1.999^n \overline{\mu}(K)$$

and also

$$\overline{\mu}(K) \leq v_n \leq 1.999^n \overline{\mu}(K) .$$

Comparing the lower bound in the first inequality with the upper bound in the second, we find that

$$1.999^{-n} v_n \leq \text{vol}(K) \leq v(n,p)$$

by the definition of $v(n,p)$. I do not know the exact value of $v(n,p)$, but to obtain the theorem, the following rough estimate will do:

$$v(n,p) \leq p \cdot 2^{-n} v_n . \tag{2.4.5}$$

Since p is polynomial in n, this is a contradiction if n is large enough.

To see (2.4.5), it suffices to notice that K is contained in the union of the "Thales–balls" over the segments connecting 0 to v_i $(i = 1, \ldots, p)$, and that the volume of each of these balls is $2^{-n}v_n$. □

If K is a polytope, then there may be much better ways to compute vol (K), but very little seems to be known in this direction. For example, if K is given as the solution set of a system of linear inequalities or as the convex hull of a set of vectors, can vol(K) be computed in polynomial time? I suspect that the answer is no, but no NP–hardness results seem to be known.

It may be worthwhile to mention here the following observation. Let $P = \{v_1, \ldots, v_n\}$ be a partially ordered set and let $K \subseteq \mathbb{R}^n$ be defined by

$$K = \{x \in \mathbb{R}^n : 0 \leq x_i \leq 1,\ x_i \leq x_j \text{ for } v_i < v_j\} .$$

Then $n!$ vol(K) gives the number of linear extensions of P. While I do not know if this number is NP–hard or not, I suspect that it is (probably even $\#P$–complete) and this may not be too difficult to prove.

Next we consider the width and the inscribed ball. Let $w(K)$ denote the width of K and let $r(K)$ denote the radius of the largest ball in K. From Theorem (2.4.1) we obtain immediately:

(2.4.6) Corollary. *For every convex body K, one can compute in polynomial time a number $w_0(K)$ such that*

$$\frac{1}{(n+1)\sqrt{n}} w_0(K) \leq 2r(K) \leq w(K) \leq w_0(K) . \qquad \square$$

An argument similar to the proof of Theorem (2.4.4) shows that one cannot get closer to $r(K)$ and to $w(K)$ than a factor of $\sqrt{n}/\log n$ in polynomial time.

The situation is similar for the smallest circumscribed ball and the diameter. Let $d(K)$ denote the diameter of K and $R(K)$ the radius of the smallest circumscribed ball.

(2.4.7) Corollary. *For every convex body K, one can compute in polynomial time a number $d_0(K)$ such that*

$$d_0(K) \leq d(K) \leq 2R(K) \leq (n+1)\sqrt{n} d_0(K) . \qquad \square$$

2.5. Integer programming in a fixed dimension.

Integer linear programming, that is, the problem of maximizing a linear objective function over the integer solutions of a system of linear inequalities, occurs in many models in operations research. In the next chapter we shall see that it also provides an important and useful framework for most combinatorial optimization problems.

It is, however, in a sense too general: it is NP–hard. In the next chapter we shall see how special cases of it, corresponding to some combinatorial optimization problems, can be solved in polynomial time. Here we treat another polynomially solvable special case, namely the case of fixed dimensional integer linear programs. We shall prove the celebrated result of H. W. Lenstra that gives a polynomial–time algorithm for this problem (where the polynomial in the time bound depends, of course, on the dimension). Lenstra introduced basis reduction techniques to solve the problem. We shall show that replacing these by the more powerful techniques described in Chapter 1, and combining them with the "shallow cut" version of the Ellipsoid Method, one can do most of the work in time which is polynomial even for varying dimension. Exponentiality enters only at a trivial case distinction at the end of the argument.

The main result of this section is the following. Recall that a *convex body* is a convex set in $\mathbb{R}^n$ given by a well–guaranteed oracle for weak separation, violation, validity or optimization, or by a centered weak membership oracle. (Since all these oracles are polynomially equivalent, it does not matter which one we choose.) The general problem is to decide if $\mathcal{K}$ contains an integer vector. The following result shows that this is the case unless $\mathcal{K}$ is "flat" in one direction.

(2.5.1) Theorem. *Given a convex body $\mathcal{K}$, we can achieve in polynomial time one of the following:*

(i) *find an integral vector in $\mathcal{K}$;*

(ii) *find an integral vector $c \in \mathbb{Z}^n$ such that*

$$\max\{c^T x : x \in \mathcal{K}\} - \min\{c^T x : x \in \mathcal{K}\} \leq 2n^2 9^n .$$

Proof. First we use Theorem (2.4.1) to find a weak Löwner–John ellipsoid pair for $\mathcal{K}$, i.e. a pair (E', E) or concentric ellipsoids such that $E' \subseteq \mathcal{K} \subseteq E$ and E' arises from E by shrinking by a factor $1/(n+1)\sqrt{n}$. Let x be the common center of E and E'.

Let τ be any linear transformation of $\mathbb{R}^n$ which maps E' onto a unit ball. Let $L = \tau(\mathbb{Z}^n)$ be the image of the standard lattice $\mathbb{Z}^n$ under this transformation, and let $y = \tau(x)$.

Now, use the algorithm of Theorem (1.2.22) to find a lattice point "near" y. This algorithm gives us a lattice point $b \in L$ as well as a proof that no lattice point is "much closer". This is supplied in terms of a vector $d \in L^*$ such that the distance of y from the two lattice hyperplanes $d^T x = \lfloor d^T y \rfloor$ and $d^T x = \lceil d^T y \rceil$ is at least $(\sqrt{2}/3)^n ||b - y||$. What we need from this is that the distance of these two lattice hyperplanes is at least $(\sqrt{2}/3)^n ||b - y||$. So $||d|| \leq (3/\sqrt{2})^n ||b - y||$.

Let $a = \tau^{-1} b$ and $c = \tau^* d$, where τ^* is the adjoint of τ.

Case 1. If $||b - y|| \leq 1$ then $b \in S(y, 1) = \tau(E')$ and hence $a \in E' \subseteq \mathcal{K}$. So we have found an integral vector in $\mathcal{K}$.

Case 2. If $||b-y|| > 1$ then $||d|| < 9^n$. Hence

$$\begin{aligned}
&\max\{c^Tx : x \in \mathcal{K}\} - \min\{c^Tx : x \in \mathcal{K}\} \\
&\quad \le \max\{c^Tx : x \in E\} - \min\{c^Tx : x \in E\} \\
&\quad = 2\max\{c^T(x-x_0) : x \in E\} \\
&\quad = 2\max\{d^T(y-y_0) : y \in \tau(E) = S(y_0,(n+1)\sqrt{n})\} \\
&\quad \le 2||d|| \cdot (n+1)\sqrt{n} < 2(n+1)\sqrt{n} \cdot (3/\sqrt{2})^n .
\end{aligned}$$

□

This result can be used to design a recursive algorithm to find a lattice vector in $\mathcal{K}$. The idea is the following. If we run the algorithm of Theorem (2.5.1) and it ends up with (i), we are done. Suppose that it ends up with (ii). Then we consider the $(n-1)$–dimensional convex bodies $\mathcal{K}_j = \mathcal{K} \cap \{x : d^Tx = j\}$ for all $j \in \mathbb{Z}$ with $\min\{c^Tx : x \in \mathcal{K}\} \le j \le \max\{c^Tx : x \in \mathcal{K}\}$. There are exponentially many cases to consider; for fixed n , however, there is only a fixed number of them, and we can proceed by induction.

A little care has to be exercised, however, with the values $j \sim \min\{c^Tx : x \in \mathcal{K}\}$ and $j \sim \max\{c^Tx : x \in \mathcal{K}\}$ (here $\sim$ means that we cannot distinguish the two values in polynomial time). Since $\mathcal{K}$ is only given by a weak oracle, we cannot really know if the hyperplanes $c^Tx = j$ avoid, touch, or intersect $\mathcal{K}$. So we can only detect those integral vectors in $\mathcal{K}$ which are in $S(\mathcal{K};\epsilon)$ for any fixed ϵ . So we have proved:

(2.5.2) Corollary. *For every fixed n , given a convex body $\mathcal{K} \subseteq \mathbb{R}^n$, we can achieve in polynomial time one of the following:*

(i) *find an integral vector in $S(\mathcal{K},\epsilon)$;*

(ii) *assert that $S(\mathcal{K},-\epsilon)$ contains no integral points.* □

Using the notion of rational polyhedra, and the techniques developed in the preceding section, it is easy to obtain a "strong" version of this theorem.

(2.5.3) Corollary. *If $P \subseteq \mathbb{R}^n$ is a polyhedron given by a strong separation (violation, optimization) oracle with rationality guarantee, then there is an algorithm which finds an integer point in P , or concludes that none exists, in time which is polynomial if the dimension n is fixed.* □

The reduction to the bounded case is based on a result of Borosh and Treybig (1976), which implies that if a rational polyhedron facet complexity ϕ contains any integral vector at all, then it contains one with $||z||_\infty < n2^{n\phi}$. The reduction to the full–dimensional case, and the refinement of the "weak" to the "strong" result can be achieved by the same methods as in the case of other algorithmic problems on rational polyhedra.

We conclude with the result of H. W. Lenstra:

(2.5.4) Corollary. *Given a system of linear inequalities in n variables, there is an algorithm which finds an integral solution, or concludes that none exists, in time which is polynomial for each fixed n .* □

CHAPTER 3

Some Applications in Combinatorics

3.1. Cuts and joins.

Perhaps the single most important combinatorial optimization problem is the flow problem. We are given a digraph G, two points $s,t \in V(G)$ (called the *source* and the *sink*), and a non-negative *capacity* $c(e)$ assigned to each edge e. To avoid some trivial difficulties, let us assume that no edge enters s and no edge leaves t. We would like to send as much "flow" from s to t as possible. The way to model this is to look for an "intensity" $f(e)$ for each edge e, so that the following conditions are satisfied:

$$0 \le f(e) \le c(e) \quad \text{for each edge } e, \tag{3.1.1}$$

and

$$\sum_{\substack{e \\ v \text{ head of } e}} f(e) = \sum_{\substack{e \\ v \text{ tail of } e}} f(e) = 0 \text{ for each point } v \ne s,t\,. \tag{3.1.2}$$

These last conditions, versions of Kirchhoff's Current Law, express that the amount of flow entering point v is the same as the amount leaving it. An assignment $f : E(G) \to \mathbb{R}$ satisfying (3.1.1) and (3.1.2) is called an (s,t)-*flow* or briefly, a *flow*. The sum on the left hand side of (3.1.2) for $v = t$ defines the *value* of the flow:

$$v(f) = \sum_{\substack{e \\ t \text{ head of } e}} f(e)\,. \tag{3.1.3}$$

It is easy to see that the same value, but with opposite sign, occurs if we consider the source point s:

$$v(f) = -\sum_{\substack{e \\ s \text{ tail of } e}} f(e)\,. \tag{3.1.4}$$

We are trying to find the maximum of $v(f)$, subject to (3.1.1), (3.1.2), (3.1.3) and (if you like) (3.1.4). This is just a linear program in the variables $f(e)$ $(e \in E(G))$ and $v(f)$. (We could easily eliminate the variable $v(f)$ from the program, and view the right hand side of (3.1.3) as the objective function. But we have chosen this form to preserve a greater degree of symmetry among the points. For the same reason, we have included the superflows equation (3.1.4) among the constraints.)

Now the maximum of $v(f)$ subject to (3.1.1) - (3.1.4) can be determined in polynomial time by various linear programming algorithms, e.g. by the Ellipsoid Method (but for this special case, the Simplex Method can also be implemented in polynomial time). Ford and Fulkerson (1962) developed an extremely important augmenting path technique to solve this problem more efficiently. But our concern is to derive its polynomial solvability from general principles.

There is also a famous min–max result for this maximum, the Max–Flow–Min–Cut Theorem of Ford and Fulkerson (1962). For a set $A \subseteq V(G)$, $s \in A$, $t \neq A$, we define the *(s,t)–cut determined by A* as the set of edges whose tail is in A and whose head is in $V(G) - A$. The *capacity* of an (s,t)–cut C is defined by $c(D) = \sum_{e \in C} c(e)$. With this terminology, we can state the following fundamental result.

(3.1.5) Max–Flow–Min–Cut Theorem. *The maximum value of an (s,t)–flow is equal to the minimum capacity of an (s,t)–cut.*

The minimum capacity of an (s,t)–cut can again be formulated as the optimum of a linear program. In fact (not too surprisingly) it is just the minimum in the dual program of (3.1.1) - (3.1.4). But this is not entirely straightforward. In the dual linear program, we will have a variable y_e for each $e \in E(G)$ corresponding to the upper bound in (3.1.1) as well as a variable z_v for each $v \in V$ corresponding to the Kirchhoff equations and the formulas for $v(f)$. This dual program is as follows:

(3.1.6) $$y_e \geq 0 \quad \text{for all} \quad e \in E(G) \ ,$$

(3.1.7) $$z_v - z_u + y_{uv} \geq 0 \quad \text{for all} \quad uv \in E(G)$$

and

(3.1.8) $$z_s - z_t = 1 \ .$$

The dual objective function is

(3.1.9) $$\text{minimize} \sum_{c \in E(G)} c(e) y_e \ .$$

Since only differences of the z's occur, we may assume that $z_t = 0, \ z_s = 1$. We can construct a solution of this program by taking an (s,t)–cut C determined by a set $A \subseteq V(G)$ $(s \in A \subseteq V(G) - t)$, and setting

$$z_u = \begin{cases} 1 & \text{if } u \in A\,, \\ 0 & \text{if } u \in V(G) - A\,; \end{cases}$$

(3.1.10)

$$y_e = \begin{cases} 1 & \text{if } e \in C\,, \\ 0 & \text{if } e \neq C\,. \end{cases}$$

Then $\sum_e c(e) y_e = \sum_{e \in C} c(e)$ is just the capacity of the cut C . So indeed the capacity of every (s,t)–cut is a value of the dual objective function (and hence, trivially, an upper bound on the maximum value is primal). But more can be said. The solutions given by (3.1.10) are not just any odd solutions of the dual program: they are precisely the vertices of the polyhedron formed by all dual solutions. This important non–trivial fact follows from the observation that the matrix of this program is *totally unimodular*, i.e. every subdeterminant of this matrix is 0 or ± 1 . So if we compute the coordinates of any vertex by Cramer's Rule, they will turn out to be integral. It takes a little play with the z_u's to show that every vertex has not only integral but 0–1 entries. But then it is obvious that every vertex is of the form given above.

If the feasible polyhedron of a linear program has any vertices, then the optimum of any (bounded) linear objective function is attained at one of these. So the optimum value of the dual program (3.1.6) – (3.1.9) is just the minimum capacity of an (s,t)–cut. Moreover, having found an optimum vertex solution to this dual program, we automatically find a minimum cut as well.

The fact that we could find a combinatorial object (a cut) by linear programming depended on the integrality of the vertices of the dual program, which in turn depended on the total unimodularity of the matrix of the program. The transpose of this matrix is also totally unimodular, and hence we obtain another important result on flows:

(3.1.11) Integrality Theorem. *If all capacities $c(e)$ are integral, then there exists an optimum flow in which all intensities $f(e)$ are integral.*

If we try to extend the Integrality Theorem to more general combinatorial situations, then the first idea is to look for other totally unimodular matrices. This approach does not lead too far, however; a very deep result of Seymour (1980) says that every totally unimodular matrix can be "glued together" (in a somewhat complicated but precisely defined sense) from matrices corresponding to the maximum flow problem or to its dual. While one does get totally unimodular linear programs which are substantially different from maximum flows (minimum cost flows, for example), the range of this approach is limited.

We can extend these methods much further if we analyse the combinatorial meaning of the Integrality Theorem. Let f be any (s,t)–flow with $v(f) > 0$. It is easy to find a directed (s,t)–path P such that $f(e) > 0$ for all $e \in P$. Let χ^P denote the incidence vector of P ; then χ^P is a special kind of (s,t)–flow with value 1. Set $\lambda_P = \min\{f(e) : e \in P\}$, then $f' = f - \lambda_P \cdot \chi^P$ is an (s,t)–flow with value $v(f') = v(f) - \lambda_P$. Going on similarly, we can decompose f into directed (s,t)–paths and possibly into a flow with value 0:

$$f = \lambda_{P_1}\chi^{P_1} + \ldots + \lambda_{P_m}\chi^{P_m} + f_0 ,$$

where $v(f_0) = 0$. So $v(f) = \lambda_{P_1} + \ldots + \lambda_{P_m}$. Now if $f(e)$ is integral then so are the λ_P , and we may view this decomposition as a collection of (s,t)–paths, in which P_i occurs with multiplicity λ_{P_i} . The condition that $f(e) \leq c(e)$ can then be translated into the condition that this family of paths uses each edge e at most $c(e)$ times. The objective function is then just the number of paths in the collection.

If we take each capacity $c(e) = 1$, then the Max–Flow–Min–Cut Theorem translates into a classical result:

(3.1.12) Menger's Theorem. *If G is a digraph and $s,t \in V(G)$, then the maximum number of edge–disjoint directed (s,t)–paths is equal to the minimum cardinality of an (s,t)–cut.*

Let us play a little with these problems by interchanging the roles of paths and cuts. Then the values $c(e)$ may be viewed as "lengths", and instead of a minimum capacity cut, we may look for a shortest directed path from s to t . Instead of packing paths, we shall be packing (s,t)–cuts. An interpretation of such a packing is not as straightforward as the "flow" interpretation of a packing of paths. But we can do the following. Let $C_1, \ldots, C_m$ be (s,t)–cuts determined by the sets $A_1, \ldots, A_m$, respectively, and $\lambda_1, \ldots, \lambda_m \geq 0$ such that $(C_1, \ldots, C_m;\ \lambda_1, \ldots, \lambda_m)$ is a c–packing of (s,t)–cuts, i.e.

$$\sum_{i_e \in C_i} \lambda_i \leq c(e)$$

holds for all $e \in E$. Consider the following function defined on the points:

$$\pi = \sum \lambda_i \chi^{E-A_i} .$$

Then $\pi(s) = 0,\ \pi(t) = \sum_i \lambda_i$, and for every edge uv we have

$$\begin{aligned} \pi(v) - \pi(u) &= \sum(\lambda_i : u \in A_i,\ v \notin A_i) \\ &\qquad - \sum\{\lambda_i : v \in A_i, u \notin A_i\} \\ &\leq \sum\{\lambda_i : u \in A_i, v \notin A_i\} \\ &= \sum_{uv \in C_i} \lambda_i \leq c(uv) . \end{aligned}$$

A function $\pi : V \to Q$ is called a *potential* (with respect to the length function c) if

$$\pi(v) - \pi(u) \leq c(uv)$$

for each edge uv . The last inequality says that to go from u to v , we have to go at least as long as the difference between the potentials of u and v .

The Max–Flow–Min–Cut Theorem has the following (substantially easier) analogue:

(3.1.13) Min–Path–Max–Potential Theorem. *The minimum length of an (s,t)–path is equal to the maximum potential difference between u and v .*

It takes the solution of a linear program to find the maximum potential difference. A basic optimum solution to the dual of this linear program would yield a shortest (s,t)–path. However, in this case there is a simple straigthforward algorithm to determine, for a fixed point s , a shortest (s,x)–path for each x , as well as a potential which simultaneously maximizes the potential difference between s and each other point x (Dijkstra (1959)).

Is there any reason why we have obtained another meaningful result by interchanging "paths" and "cuts"? Is there any connection between the algorithmic solvability of the shortest path and maximum flow problems? To answer these questions, let us formulate these problems quite generally.

Suppose that instead of the paths, we have any collection $\mathcal{H}$ of subsets of a finite set E such that no member of $\mathcal{H}$ contains another. Such a set–system is called a *clutter.* We can generalize the maximum flow problem as follows. Given "capacities" $c(e) \in \mathbb{Z}_+$ for each $e \in E$, find a maximum collection of members of $\mathcal{H}$ which use every $e \in E$ at most $c(e)$ times. We call such a collection a *c–packing* of members of $\mathcal{H}$. We can also formulate a "fractional" version of this problem: find a list $\{P_1, \dots, P_m\}$ of members of $\mathcal{H}$ and non–negative rational numbers $\lambda_1, \dots, \lambda_m$ such that

$$\sum_{P_i \ni e} \lambda_i \leq c(e) \quad (e \in E)$$

and $\sum \lambda_i$ is maximum. Such a system $\{P_1, \dots, P_m, \lambda_1, \dots, \lambda_m\}$ is called a *fractional* c–packing of members of $\mathcal{H}$.

What will correspond to cuts in this general setting? Observe that an (s,t)–cut meets every (s,t)–path; and that every set of edges meeting all (s,t)–paths contains an (s,t)–cut. It may happen that an (s,t)–cut contains another (s,t)–cut as a proper subset; but as long as we are only interested in the minimum capacity of (s,t)–cuts, we may disregard the first. So the inclusion-minimal (s,t)–cuts are just the minimal sets of edges which meet every (s,t)–path.

This motivates the following definition: the *blocker* $\mathrm{bl}(\mathcal{H})$ of $\mathcal{H}$ consists of all minimal sets which meet every member of $\mathcal{H}$. If $(P_1, \dots, P_m; \lambda_1, \dots, \lambda_m)$ is any (integral or fractional) c–packing of members of $\mathcal{H}$, and if $Q \in \mathrm{bl}(\mathcal{H})$, then

$$\sum_{i=1}^{m} \lambda_i \leq \sum_{i=1}^{m} \lambda_i \, |P_i \cap Q| = \sum_{e \in Q} \sum_{P_1 \ni e} \lambda_i \leq \sum_{e \in Q} c(e) = e(Q) \, .$$

So the minimum capacity of members of $\mathrm{bl}(\mathcal{H})$ is an upper bound on the maximum value of a fractional c–packing of members of $\mathcal{H}$.

We say that $\mathcal{H}$ has the *max-flow-min-cut property* if equality holds here for every capacity function $c : E \to \mathbb{Q}^+$. It is a common form of very many combinatorial min–max results that certain clutters $\mathcal{H}$ have the max-flow-min-cut property.

It is easy to verify that $\mathrm{bl}(\mathrm{bl}(\mathcal{H})) = \mathcal{H}$ for every $\mathcal{H}$. A simple but important result of Lehman (1965) asserts that if $\mathcal{H}$ has the max-flow-min-cut property then so does $\mathrm{bl}(\mathcal{H})$. So in a sense the Max-Flow-Min-Cut Theorem and the Min-Path-Max-Potential Theorem are equivalent: they follow from each other by Lehman's result. This result will be better understood (and quite simply proved) if we introduce some further polyhedral considerations.

Let $\mathcal{H}$ be any clutter of subsets of a set E . Define the *dominant* of $\mathcal{H}$ by

$$\mathrm{dmt}\mathcal{H} = \mathrm{conv}\{\chi^P : P \in \mathcal{H}\} + \mathbb{R}_+^E .$$

If we have any function $c : E \to Q_+$, and want to find a set $P \in \mathcal{H}$ with $c(P)$ minimum (e.g. a minimum capacity (s,t)-cut or a shortest (s,t)-path), then we can write this as

$$\text{(3.1.14)} \qquad \text{minimize } \{c^T x : x \in \mathrm{dmt}\mathcal{H}\} .$$

In fact, this minimum is finite and will therefore be achieved by a vertex of $\mathrm{dmt}\mathcal{H}$. But the vertices of $\mathrm{dmt}\mathcal{H}$ are, trivially, the vectors χ^P, $P \in \mathcal{H}$, and $c^T\chi^P = c(P)$ for such vectors.

So we have formulated the problem of finding a minimum weight member of $\mathcal{H}$ as the problem of minimizing a linear objective function over a polyhedron. This is just a linear program! Well, not quite ... . To be able to handle it as a linear program, we would need a description of $\mathrm{dmt}\,\mathcal{H}$ as the solution set of a system of linear inequalities. It is well known that such a description does exist for each polyhedron, but how to find it in our case?

Let us guess at some inequalities which are at least valid for all vectors in $\mathrm{dmt}\,\mathcal{H}$. The class

$$\text{(3.1.15)} \qquad x_e \ge 0 \quad \text{for all } e \in E$$

is trivial. For any $Q \in \mathrm{bl}(\mathcal{H})$ we obtain a non-trivial constraint:

$$x(Q) \ge 1 \quad (Q \in \mathrm{bl}\,(\mathcal{H})) .$$

If $x = \chi^P$ is any vertex of $\mathrm{dmt}\,\mathcal{H}$ then

$$\text{(3.1.16)} \qquad x(Q) = \sum_{e \in Q} x_e = |P \cap Q| \ge 1 ,$$

and this inequality is trivially inherited for all vectors in the convex hull and "above".

Now the max–flow–min–cut property for bl ($\mathcal{H}$) holds if and only if the inequalities (3.1.15) and (3.1.16) suffice to describe that a fractional packing of members of bl ($\mathcal{H}$) is just a feasible dual solution to the linear program

$$\begin{aligned}&\text{maximize } c^T x\\&\text{subject to (3.1.15) and (3.1.16)}\ ,\end{aligned}$$

and that this has the same optimum value as (3.1.14) if and only if (3.1.15) and (3.1.16) describe exactly dmt $\mathcal{H}$.

We may find yet another equivalent form of this property.

The polyhedron dmt $\mathcal{H}$ has the following property: whenever $x \in$ dmt $\mathcal{H}$ and $y \geq x$ then also $y \in$ dmt $\mathcal{H}$. A set with this property will be called *up–monotone.*

Let $K \subseteq \mathbb{R}_+^E$ be any up–monotone convex polyhedron. We define the *blocking polyhedron* BL(K) of K by

$$\mathrm{BL}(K) = \{y \in \mathbb{R}_+^E : x^T y \geq 1 \ \text{ for all } \ x \in K\}.$$

This notion, which is a version of the classical notion of polarity of polyhedra, is due to Fulkerson (1970). It follows by standard linear algebra that BL(K) is again up–monotone and

$$\mathrm{BL}(\mathrm{BL}(K)) = K\ .$$

There are a number of nice relations between K and BL(K) , for example, the vertices of K correspond nicely to the non–trivial facets of BL(K) and vice versa, but we shall not elaborate on these.

Let us determine the blocking polyhedron of dmt $\mathcal{H}$. By definition,

$$\mathrm{BL}(\mathrm{dmt}\mathcal{H}) = \{y \in \mathbb{R}_+^E : x^T y \geq 1 \ \text{ for all } x \in \ \mathrm{dmt}\mathcal{H}\}\ .$$

Since an inequality $x^T y \geq 1$ ($y \geq 0$) holds true for all $x \in$ dmt$\mathcal{H}$ if and only if it holds true for each of the vertices of dmt $\mathcal{H}$, we can write this as

$$\mathrm{BL}(\mathrm{dmt}\mathcal{H}) = \{y \in \mathbb{R}_+^E : y^T \chi^P = y(P) \geq 1 \ \text{ for all } P \in \mathcal{H}\}\ .$$

Interchanging the role of $\mathcal{H}$ and bl $\mathcal{H}$ we obtain

$$\mathrm{BL}(\mathrm{dmt}(\mathrm{bl}(\mathcal{H}))) = \{y \in \mathbb{R}_+^E : y(Q) \geq 1 \text{ for all } Q \in \ \mathrm{bl}(\mathcal{H})\}\ .$$

The right-hand side here is just the solution set of (3.1.15) and (3.1.16). So $\mathcal{H}$ has the max–flow–min–cut property if and only if

$$\mathrm{BL}(\mathrm{dmt\ bl}(\mathcal{H})) = \ \mathrm{dmt}\mathcal{H}\ .$$

Applying the BL operator to both sides we get

$$\mathrm{dmt\ bl}(\mathcal{H}) = \ \mathrm{BL}(\mathrm{dmt}(\mathcal{H}))\ .$$

But this simply means that $\mathrm{bl}(\mathcal{H})$ has the max–flow–min–cut property! So we have proved Lehman's Theorem.

Let us complement this by an algorithmic result.

Note that it is easy to give rationality guarantees for $\mathrm{dmt}\,(\mathcal{H})$. So if $\mathcal{H}$ has the max–flow–min–cut property, then by the results of Chapter 2, the Optimization Problem for $\mathrm{dmt}\mathcal{H}$ is polynomial time equivalent to the Optimization Problem for $\mathrm{dmt}(\mathrm{bl}(\mathcal{H}))$. This yields the following general result.

(3.1.17) Theorem. *Let $\mathcal{K}$ be a class of clutters $(E, \mathcal{H})$ with the max–flow–min–cut property, encoded so that the problem "Given $(E, \mathcal{H}) \in \mathcal{K}$ and $c : E \to \mathbb{Q}_+$, find $P \in \mathcal{H}$ with $c(P)$ minimum" can be solved in polynomial time. Then the problem "Given $(E, \mathcal{H}) \in \mathcal{K}$ and $c : E \to \mathbb{Q}_+$, find $Q \in \mathrm{bl}(\mathcal{H})$ with $c(Q)$ minimum" can be solved in polynomial time.*

In particular, the fact that a shortest (s,t)–path can be found in polynomial time implies in this generality that a minimum capacity (s,t)–cut can be found in polynomial time. In what follows, we shall show that Theorem (3.1.17) applies to many other situations.

But first let me make some comments on the conditions of this theorem. First, the hypothesis that the clutters have the max–flow–min–cut property is essential. For example, if $(E, \mathcal{H})$ is a graph explicitly given (i.e., if $\mathcal{H}$ consists of some 2–element subsets of E), then to find $\min\{c(P) : P \in \mathcal{H}\}$ is trivial if we scan all edges. On the other hand, $\mathrm{bl}(\mathcal{H})$ consists of all point–covers, and to find a minimum weight (even minimum cardinality) point–cover is NP–hard.

Second, the hypothesis that a minimum weight member of $\mathcal{H}$ can be chosen in polynomial time may be superfluous, as far as I know. It may perhaps be true that if all the members of $\mathcal{K}$ are encoded in a reasonable way, and if they all have the max–flow–min–cut property, then a minimum weight member of $\mathcal{H}$ can be found in polynomial time. There is a somewhat similar result for the "antiblocking" situation, to be discussed in the next section.

Now let us describe other examples of hypergraphs with the max–flow–min–cut property and the algorithmic implications of Theorem (3.1.17). Various families of cuts give rise to such hypergraphs.

Let G be a digraph and $s \in V(G)$. An *s–cut* is a cut determined by a set $A \subset V(G)$ such that $s \in A$. So the set of s–cuts is the union of the sets of (s,t)–cuts for $t \in V(G) - S$.

It is easy to work out that the blocker of s–cuts consists of all spanning trees of G directed so that every point different from s has indegree 1. Such a spanning tree will be called an *s–branching.* The clutter of s–cuts (and therefore, by Lehman's Theorem, the clutter of s–branchings) has the max–flow–min–cut property. These two results were proved by Edmonds (1973) and Fulkerson (1974). (The results of Edmonds and Fulkerson assert in fact more; they also imply the existence of integral optimum dual solutions. We shall return to this later.)

It is trivial to find a minimum weight s–cut in polynomial time: all we have to do is to find a minimum weight (s,t)–cut for each $t \in V(G) - s$, and pick the best of these. Theorem (3.1.17) implies then that we can also find a minimum weight s–branching in polynomial time. This task is non–trivial; a direct polynomial algorithm to solve it was given by Edmonds (1967).

A next possible step would be to consider all cuts, but this clutter does not have the max–flow–min–cut property. Its blocker consists of all strongly connected subgraphs of G. The problem of finding a minimal strongly connected subgraph contains the Hamilton circuit problem, and is, therefore, NP–hard.

But we get another nice family if we consider directed cuts or, briefly, dicuts. A *dicut* is a cut determined by a set $A \subset V(G)$, $A \neq \emptyset$ if there is no edge with tail in $V(G) - A$ and head in A. The blocker of the dicuts will consist of all minimal sets $P \subseteq E(G)$ such that by contracting P we obtain a strongly connected digraph. Such a set will be called a *dijoin.* The clutter of dijoins (as well as the clutter of dicuts) has the max–flow–min–cut property. This follows from a theorem of Lucchesi and Younger (1978).

It is not difficult to find a minimum capacity dicut in a graph. Let us add, for each edge uv, a reverse edge uv with infinite capacity. Then every cut will have infinite capacity, except the dicuts, and so a minimum capacity cut in these modified digraphs is just a minimum capacity dicut in the original.

By Theorem (3.1.17), we can find then a minimum length dijoin in polynomial time. Again, this is quite a difficult task, and the direct algorithms to achieve it (Lucchesi (1976), Frank (1981)) are quite involved.

Perhaps the most interesting families of cuts and joins with the max–flow–min–cut property are T–cuts and T–joins. Let G be an undirected graph and $T \subseteq V(G)$, $|T|$ even. A *T–cut* is a cut determined by a set $A \subseteq V(G)$ with $|A \cap T|$ odd. A *T–join* is a minimal set of edges meeting every T–cut. It is easy to work out what T–joins are: they are those subforests of G which have odd degree at the points of T and even degree (possibly 0) elsewhere. So if $T = \{u, v\}$ then the T–cuts are the (u, v)–cuts and the T–joins are the (u, v)–paths.

As another special case of interest, let $T = V(G)$. Then T–cuts are cuts with an odd number of points on both sides, while T–joins are spanning forests with all degrees odd. In particular, every T–join has at least $\frac{1}{2}|V(G)|$ edges, and equality holds if the T–join is a perfect matching. So to find a minimum cardinality T–join contains the problem of finding a perfect matching. The problem of finding a minimum weight perfect matching (one of the most prolific problems in discrete optimization) can also be reduced to a minimum weight T–join problem. Let G be a graph with at least one perfect matching and $w : E(G) \to \mathbb{Q}_+$ any weighting of its edges. Add a big constant N to all weights, then the weight of any T–join which is not a perfect matching is automatically larger than the weight of any perfect matching. So if we look for a minimum weight T–join using these modified weights, then only perfect matchings come into consideration. But for perfect matchings, adding this constant N to all edges means adding $\frac{1}{2}N \cdot |V(G)|$ to the weight of each perfect matching, and so finding one with minimum new weight is tantamount to finding one with minimum old weight.

Let us remark that, conversely, the problem of finding a minimum weight T-join can be reduced to finding a minimum weight perfect matching in some auxiliary graph; this is the way Edmonds and Johnson (1973) solved the problem.

It follows from the results of Edmonds and Johnson that the clutter of T-joins (and the clutter of T-cuts) has the max-flow-min-cut property. Hence Theorem (3.1.17) implies that the problem of finding a minimum weight T-join is polynomial time equivalent to the problem of finding a minimum weight T-cut. Neither one of these problems is obvious; yet a minimum weight T-cut can be found by a conceptually simpler and more general procedure, which also reduces the problem to several minimum cut computations. From this the polynomial time solvability of the minimum T-join problem (and hence of the weighted matching problem) follows by general principle.

Let us describe an algorithm to find a minimum weight T-cut, due to Padberg and Rao (1982). First, we find a minimum weight cut C separating at least two points in T. This can be done by the usual trick, finding a minimum (s,t)-cut for each pair $s,t \in T$ of points and taking the best of these. Now if C happens to be a T-cut, then it is clearly minimal, and we are done.

So suppose that C is not a T-cut. Then C is determined by a set A such that $|T \cap A|$ is even. Since C separates at least two points of T, we have that $|T \cap A| \geq 2$ and $|T - A| \geq 2$.

Construct two new graphs G_1 and G_2 by contracting the sets A and $V(G) - A$, respectively, to a single point. Keep the old weights on all the non-contracted edges. Set $T_1 = T - A$, $T_2 = T \cap A$. Find a minimum weight T_i-cut C_i in G_i $(i = 1,2)$ and let, say, C_1 have the smaller weight of these two cuts. C_1 may also be viewed as a T-cut in G. We claim that C_1 is a minimum weight T-cut in G.

To this end, let C' be any other T-cut in G, determined by a set $B \subseteq V(G)$. Then $|T \cap B|$ is odd and hence one of $|T \cap B \cap A|$ and $|T \cap B - A|$ is odd, the other is even. Suppose e.g. that $|T \cap B \cap A|$ is odd.

Since $T - A \neq \emptyset$ it follows that either $(T-A) \cap B$ or $T - A - B$ is non-empty. We may assume that $T - A - B \neq \emptyset$, or else we can replace B by $V(G) - B$. Now the cut C^* determined by $A \cup B$ separates T, and hence

$$w(C^*) \geq w(C)$$

by the choice of C. Moreover, the T-cut C^{**} determined by $A \cap B$ is also a T-cut in G_2, and hence

$$w(C^{**}) \geq w(C_2) \geq w(C_1) .$$

An easy calculation yields the simple but important inequality

$$w(C^*) + w(C^{**}) \leq w(C) + w(C') .$$

Hence

$$\begin{aligned} w(C') &\geq w(C^*) + w(C^{**}) - w(C) \\ &\geq w(C) + w(C_1) - w(C) = w(C_1) . \end{aligned}$$

Thus C_1 is indeed a minimum weight T–cut in the whole graph.

This consideration gives rise to a recursive algorithm to compute a minimum weight T–cut. With a little care it is not difficult to see that its running time is polynomial.

Let $\mathcal{H}$ be a clutter with the max–flow–min–cut property. Along with the problem of finding a minimum weight member of $\mathcal{H}$ goes a dual problem: find a maximum fractional packing of members of bl $(\mathcal{H})$. Quite often it happens that this fractional packing problem has an integral optimum solution. In this case we say that bl $(\mathcal{H})$ has the $\mathbb{Z}_+$*–max–flow–min–cut property.* The clutters of (s,t)–paths, (s,t)–cuts, s–branchings, s–cuts and directed cuts have the $\mathbb{Z}_+$–max–flow–min–cut property; but the clutters of dijoins, T–cuts and T–joins do not. So this property is not preserved by the blocker operation.

If we can find a minimum weight member of bl $(\mathcal{H})$ (or, equivalently, of $\mathcal{H}$) in polynomial time, then we can optimize any linear objective function over dmt bl $(\mathcal{H})$. As we remarked in Chapter 2, this also implies that we can find a dual solution in polynomial time, i.e. a maximum fractional packing of members of $\mathcal{H}$. We may even find a basic dual solution.

Assume now that $\mathcal{H}$ has the $\mathbb{Z}_+$–max–flow–min–cut property. Does this mean that we can find an optimum *integral* packing of members of $\mathcal{H}$? In the case of (s,t)–paths and (s,t)–cuts, the total unimodularity implies that the basic dual solutions are automatically integral. So if we find a basic dual solution we have also found an optimum integral packing of members of $\mathcal{H}$.

This does not work as simply in the case of s–branchings, s–cuts and dicuts; one needs special techniques to turn a basic, but possibly non–integral, optimal dual solution into an integral one. I do not know if there is a general polynomial time algorithm to do so.

3.2. Chromatic number, cliques, and perfect graphs.

The chromatic number is one of the most famous (or notorious) invariants of a graph; it suffices to refer to the Four Color Theorem. Recall that the chromatic number is defined as the least number k for which the nodes of the graph can be k–colored so that the two endpoints of any edge have different colors. A set of nodes no two of which are adjacent is called *stable.* So a coloration is a partition of $V(G)$ into stable sets. We denote the chromatic number of G by $\chi(G)$.

To determine the chromatic number of a general graph is a very difficult (NP–hard) problem. One general approach is to find bounds on this number. For our purposes, it will suffice to consider only lower bounds. We start with one which is perhaps the most trivial. Let $\omega(G)$ denote the number of nodes in a largest complete subgraph of G . Since these $\omega(G)$ points must be colored differently in every legal coloration, we have

$$\omega(G) \leq \chi(G) . \tag{3.2.1}$$

The example of a pentagon shows that equality does not always hold here. The graphs for which equality holds are interesting because for them this common value of $\omega(G)$ and $\chi(G)$ can be determined in polynomial time, as we shall see below. In fact, we can prove the following (Grötschel, Lovász and Schrijver (1981)).

(3.2.2) Theorem. *One can compute in polynomial time a number $f_0(G)$ such that*

$$\omega(G) \le f_0(G) \le \chi(G) .$$

Before proving this theorem, we generalize it in a certain way. Such a generalization will be essential to the algorithm, at least if we also want to find an optimum coloration (not merely the number of colors in one).

Let $w : V(G) \to \mathbb{Z}_+$ be any weighting of the nodes of a graph G with non-negative integers. Then we denote by $\omega(G; w)$ the maximum weight of a complete subgraph in G. We define a *w-coloration* of G as a list $(A_1, \dots, A_k)$ of stable sets and a list $(\lambda_1, \dots, \lambda_k)$ of positive integers such that $w = \lambda_1 \chi^{A_1} + \dots + \lambda_k \chi^{A_k}$. We may view λ_i as the "multiplicity" of A_i in the multiset $\{\lambda_1 \cdot A_1, \dots, \lambda_k \cdot A_k\}$. The number $\lambda_1 + \dots + \lambda_k$ is then the *number* of *colors* in this w-coloring. We denote by $\chi(G; w)$ the minimum number of colors in any w-coloring of G. Then we also have

$$\omega(g; w) \le \chi(G, w) .$$

Moreover, we shall prove

(3.2.3) Theorem. *One can compute in polynomial time a number $f_0(G; w)$ such that*

$$\omega(G; w) \le f_0(G; w) \le \chi(G; w) .$$

Remarks. 1. Obviously, if we set $w \equiv 1$ here then we obtain the preceding theorem.

2. It might sound easier to define the number $\chi(G, w)$ as the minimum number of colors in a coloration of the nodes where each node x gets not one but $w(x)$ colors, and no two adjacent nodes have any color in common. But if w is large, to encode such a coloration would take about $w(V(G))$ space rather than $\langle w \rangle$. In the cases which we will consider, we shall be able to describe an optimum coloration in polynomial space by specifying a list $(A_1, \dots, A_k)$ where $k \le n^{\text{const}}$ and then the "multiplicities" λ_i, which then clearly have $\langle \lambda_i \rangle \le \langle w \rangle$.

3. It is easy to observe that the weighted problems introduced above may be reduced to their unweighted versions by replacing each node v by a set B_v of $w(v)$ nodes, and connecting two nodes $x \in B_u$ and $y \in B_v$ if and only if $u = v$ or uv is an edge of G. This "blown up" graph G_ω has

$$\chi(G_w) = \chi(G, w), \ \omega(G_w) = \omega(G, w) .$$

Such a reduction allows one to extend several results on chromatic number and complete graphs to the weighted case. From an algorithmic point of view, however, it is not very useful, since it increases the size of the graph considerably:

$|V(G_w)| = w(V(G))$, which is not polynomial in $\langle w \rangle$. If, however, $w(V(G))$ is bounded by a polynomial in $|V(G)|$, then this reduction is polynomial.

4. It would be interesting to find other examples of "sandwich theorems" like (3.2.2), asserting that if f, g are two graph invariants such that $f \leq g$ and the properties $f \geq k$ and $g \leq k$ are in NP , then there is a polynomially computable invariant h such that $f \leq h \leq g$.

The proof of Theorem (3.2.3) depends on the right definition of $f_0(G)$. For now, I know of (essentially) one such polynomially computable function. The direct definition of this is quite *ad hoc*, and the way we shall present it is, unfortunately, quite involved. This function was introduced by Lovász (1979) as a bound on the so-called Shannon capacity of a graph. We shall describe the proof in the unweighted case, i.e. if $w \equiv 1$. For further motivation and an extension of these methods to the "weighted case", we refer to Grötschel, Lovász and Schrijver (1984a, 1984d).

So let us jump into the middle of things, and consider a graph G on $V(G) = \{1, \ldots, n\}$. Let $\mathcal{A}$ be the set of all symmetric $n \times n$ matrices for which $(A)_{ij} = 1$ if $i = j$ or if i and j are adjacent in G . We let the elements of A corresponding to non–adjacent positions vary.

For any symmetric matrix A , we denote by $\Lambda(A)$ its largest eigenvalue. (Since A is symmetric, its eigenvalues are real.) This value $\Lambda(A)$ is not necessarily the eigenvalue of A with the largest absolute value; A may have negative eigenvalues with larger absolute value.

We define

$$\vartheta(G) = \min\{\Lambda(A) : A \in \mathcal{A}\} .$$

Theorem (3.2.2) follows if we prove the following two results.

(3.2.4) Lemma. *For every graph* G , $w(G) \leq \vartheta(G) \leq \chi(G)$.

(3.2.5) Lemma. $\vartheta(G)$ *can be computed in polynomial time.*

A word of warning is in order here: the number $\vartheta(G)$ is in general not even rational! For example, one can show that if G is the pentagon then $\vartheta(G) = \sqrt{5}$. So we have to determine $\vartheta(G)$ in the sense of a real number box, i.e., for every $\epsilon > 0$ we can compute in polynomial time a rational number r such that $|r - \vartheta(G)| < \epsilon$.

To compute a function $f_0(G)$ as in Theorem (3.2.2), we choose $\epsilon = \frac{1}{3}$. Let r be the output of the algorithm in Lemma (3.2.5). Then we define $f_0(G)$ as the integer nearest to r . It follows from Lemma (3.2.4) that

$$w(G) - \frac{1}{3} \leq r \leq \chi(G) + \frac{1}{3}$$

and so

$$w(G) \leq f_0(G) \leq \chi(G)$$

as claimed.

So indeed it suffices to prove Lemmas (3.2.4) and (3.2.5).

Proof of Lemma (3.2.4). Let, say, nodes $1, 2, \ldots, w$ of G form a maximum complete subgraph of G. Then the upper left $w \times w$ submatrix A' of any $A \in \mathcal{A}$ consists of 1's, and so

$$\Lambda(A) \geq \Lambda(A') = w .$$

Hence

$$\vartheta(G) = \min \{\Lambda(A) : A \in \mathcal{A}\} \geq w(G) .$$

To prove the other inequality, let $k = \chi(G)$. It suffices to consider the case of "k–partite Turán graphs", i.e. the case where $V(G)$ has a partition $V(G) = V_1 \cup \ldots \cup V_k$ such that $|V_1| = \ldots = |V_k| = m$ (say) and two nodes are adjacent if and only if they belong to different classes V_i. Let B denote the adjacency matrix of the complement of $\overline{G}$ of G, i.e. let $(B)_{ij} = 1$ if and only if i and j are adjacent points in $\overline{G}$. Then clearly $A = J + tB$ belongs to $\mathcal{A}$ for all real t (where J is the $mk \times mk$ matrix of all 1's). Since G is a regular graph, B has constant row and column sums and hence J and B commute. Since J has eigenvalues $km, 0, \ldots, 0$ and B has eigenvalues $m - 1$ (k times) and -1 ($k(m-1)$ times), the matrix A has eigenvalues $km + t(m-1)$, $t(m-1)$ ($k - 1$ times) and $-t$ ($k(m-1)$ times). For the choice $t = -k$, largest of these is k. So for this t, $\Lambda(A) = k$ and so $\vartheta(G) \leq \Lambda(A) = k$. □

Proof of Lemma (3.2.5). By elementary properties of eigenvalues, the largest eigenvalue $\Lambda(A)$ is a convex function of the (symmetric) matrix A. This convex function must be minimized over the affine subspace $\mathcal{A}$. This can be done in polynomial time by Theorem (2.2.15), provided we can find a ball or cube about which we know *a priori* that it contains the minimum.

Suppose that $A \in \mathcal{A}$ minimizes $\Lambda(A)$. Then we can write $A = T^* DT$, where T is an orthogonal matrix and D is the diagonal matrix formed by the eigenvalues $\lambda_1, \ldots, \lambda_n$ of A. Note that $\sum_{i=1}^n \lambda_i = \operatorname{Tr} A = n$, and hence $\Lambda(A) \geq \lambda_i = n - \sum_{j \neq i} \lambda_i \geq n - (n-1)\Lambda(A)$ for all i. Since trivially $\Lambda(A) \leq n$, we have $|\lambda_i| \leq n^2$. Hence

$$|a_{ij}| = \left| \sum_{k=1}^n \lambda_k T_{ik} T_{jk} \right| \leq n^2 \sum_{k=1}^n |T_{ik}| \, |T_{jk}|$$
$$\leq n^2 \left(\sum_{k=1}^n |T_{ik}|^2 \right)^{1/2} \left(\sum_{k=1}^n |T_{jk}|^2 \right)^{1/2} = n^2 .$$

So it suffices to minimize $\Lambda(A)$ over the polytope $\{A \in \mathcal{A} : ||A||_\infty \leq n^2\}$.

We have to add one more thing: we have to be able to evaluate the function $\Lambda(A)$ in polynomial time for each rational $A \in \mathcal{A}$ with $||A||_\infty \leq n^2$, with given error $\epsilon > 0$. Almost any standard algorithm in linear algebra would do here (at least in theory), although the polynomiality of the procedures is usually not proved explicitly. Let us describe one, whose polynomiality is perhaps the easiest to verify (although it is certainly not the most efficient).

For any real number t, we have $t > \Lambda(A)$ iff $tI - \Lambda(A)$ is positive definite. So it suffices to be able to check in polynomial time whether or not a matrix is positive definite; then we can determine $\Lambda(A)$ by binary search.

Now the matrix A is positive definite if and only if all the upper left-hand corner subdeterminants (the principal minors) are positive. These n determinants can be evaluated in polynomial time. (In fact, a single Gaussian elimination procedure suffices to compute all these determinants: if we pivot at each step on the next element in the main diagonal, then the products of the first k of these elements, $k = 1, \ldots, n$, give the determinants of these principal minors. So it suffices to check whether or not these elements are all positive.) $\square$

Assume that G is a graph for which $w(G) = \chi(G)$ holds. Then of course also $f_0(G) = \vartheta(G) = w(G) = \chi(G)$, so for such graphs, the chromatic number and the size of the maximum clique can be computed in polynomial time.

Unfortunately, it is NP–complete to decide whether or not $w(G) = \chi(G)$ holds for a given graph. We can reduce the well-known NP–complete problem of deciding whether a graph H is 3–colorable to this as follows. We may assume that H does not contain a complete 4–graph, or else the answer is trivially "no". Let G be the disjoint union of H with a triangle. Then $\chi(G) = w(G)$ iff H is 3–colorable.

A similar simple construction shows that it is NP–hard to find, in every graph G with $w(G) = \chi(G)$, a maximum clique and a minimum coloration.

If we look, however, at the following subclass of such graphs, we obtain much nicer results and still retain most of the interesting examples. A graph G is called *perfect* if $w(H) = \chi(H)$ holds for every induced subgraph of G (i.e. for all subgraphs obtained by deleting some nodes and those lines adjacent to these nodes). This notion was introduced by Berge in 1960, and since then many interesting structural properties, as well as many interesting classes, of perfect graphs have been discovered. We cannot go into these questions here; the interested reader should consult Golumbic (1980) or Berge and Chvátal (1984). Let us just mention some of the most important examples of perfect graphs.

(3.2.6) *Example.* Let G be a bipartite graph. Then trivially G is perfect. But also the complement of G is perfect; this fact translates to the classical result of König that in every bipartite graph, the maximum size of a stable set of nodes is equal to the minimum number of edges covering all nodes. We may also consider the line–graph of G, i.e. the graph $\mathcal{L}(G)$ whose nodes are the edges of G; two are connected by an edge if and only if they have a node in common. This graph is also perfect; this is equivalent to a second classical result of König asserting that the chromatic index of any bipartite graph is equal to the maximum degree of its nodes. Finally, consider the complement of the line–graph of a bipartite graph. This graph is again perfect, by a third famous theorem of König. This asserts in direct terms that the maximum number of independent edges of any bipartite graph is equal to the minimum number of nodes covering all edges.

(3.2.7) *Example.* Let $(P, \leq)$ be a poset, and let G be the comparability graph of P, i.e. the graph obtained by connecting any two comparable elements of P by an edge. Then it is an easy exercise to show that G is perfect. We may also consider the complement of G. This is also perfect; this result is equivalent to Dilworth's Theorem on partitioning a poset into chains.

(3.2.8) *Example.* Let G be a *rigid circuit graph,* i.e. a graph in which every circuit longer than 3 has a chord. Then both G and its complement are perfect.

It is no coincidence that perfect graphs come in pairs in these examples. In fact, *the complement of every perfect graph is perfect.* This fact was conjectured by Berge and proved in Lovász (1972). Fulkerson (1971) was motivated by this conjecture when he developed the theory of anti–blocking polyhedra. Since these show a nice analogy with the theory of blocking polyhedra, whose combinatorial aspects were discussed in Section 3.1, and also play an important role in the algorithmic treatment of perfect graphs, we shall discuss them briefly.

For any graph G, we define the *clique polytope* of G by

$$Q(G) = \text{conv}\ \{\chi^Q : Q \subseteq V(G),\ Q \text{ spans a complete subgraph}\}\ .$$

We shall also be interested in the clique polytope of the complement of $G, Q(\overline{G})$. The vertices of $Q(\overline{G})$ are the incidence vectors of stable sets of points in G, and $Q(\overline{G})$ is often called the *stable set polytope* or *vertex packing polytope* of G.

The significance of the clique polytope lies in the fact that it contains, in a sense, the solution of the problem of finding a maximum weight clique in G for each weighting of the nodes of G. More exactly, if $w : V(G) \to \mathbb{Z}_+$ is any weighting of the nodes of G then we have

$$\omega(G; w) = \max\ \{w^T x : x \in Q(G)\}\ .$$

It would be desirable to have a description of $Q(G)$ as the solution set of a system of linear inequalities. This is of course an NP–hard task in general, but for certain special classes of graphs, in particular for perfect graphs, it has a nice answer.

It is easy to find the following inequalities, which are valid for all vectors in $Q(G)$ (although in general they are not sufficient to characterize $Q(G)$) :

$$\begin{aligned} x_v &\geq 0 \ \text{ for every } v \in V(G)\ , \\ \sum_{v \in A} x_v &\leq 1 \ \text{ for every stable set } A \subseteq V(G)\ . \end{aligned} \tag{3.2.9}$$

Now the following result shows that perfect graphs form a very well-behaved class of graphs from the point of view of polyhedral combinatorics (Fulkerson (1971), Chvátal (1975)):

(3.2.10) Theorem. *A graph* G *is perfect if and only if its clique polytope is exactly the solution set of* (3.2.8). □

We can put this result in a different form. Let $P \subseteq \mathbb{R}^n_+$ be a non–empty polyhedron. We say that P is *down–monotone* if whenever $x \in P$ and $0 \le y \le x$ then also $y \in P$. So e.g. the clique polytope of a graph is down–monotone. We define the *antiblocker* of P by

$$ABL(P) = \{x \in \mathbb{R}^n_+ : y^T x \le 1 \text{ for all } y \in P\} .$$

(The only difference from the usual notion of *polar* is that we restrict ourselves to the non–negative orthant.) It is not difficult to see that $ABL(P)$ is down–monotone. Furthermore, $ABL(ABL(P)) = P$.

Now Theorem (3.2.10) can be rephrased as follows.

(3.2.11) Corollary. *A graph* G *is perfect if and only if* $Q(G)$ *and* $Q(\overline{G})$ *are antiblockers of each other.* □

Note that in this form the corollary also contains the assertion that G is perfect iff $\overline{G}$ is perfect.

So perfect graphs lead to interesting pairs of antiblocking polyhedra with integral vertices. In fact, every pair of antiblocking polytopes with integral vertices consists of the clique polytopes of a pair of complementary graphs.

Turning our attention to algorithmic problems on perfectness, we have to start with an unsolved problem: no algorithm is known to test perfectness of a graph in polynomial time. It is not known either whether this problem is NP–hard. (In view of the positive results below, I would guess that perfectness can be tested in polynomial time.)

Suppose now that G is perfect. We can find the value of $\omega(G)$ by Lemma (3.2.5). But more than that, we can find a maximum clique in G in polynomial time. To this end, it suffices to remove points from G as long as we can do so without decreasing $\omega(G)$. The remaining points then form a maximum clique.

It is more difficult to find an optimum coloration, and we have to use a few more of the general algorithmic results on polyhedra. The fact that we can find a maximum weight clique in G in polynomial time means that the optimization problem for $Q(G)$ is polynomially solvable. Since $Q(G)$ has integral vertices, an appropriate rationality guarantee is trivial. So we can use the results of Section 2.3, in particular the assertion of Theorem (2.3.5) on problem (c). For the valid inequality $1 \cdot x \le w(G)$, we find facets $a_i^T x \le \alpha_i$ $(i = 1, \ldots, n)$ and numbers $\pi_1, \ldots, \pi_n$ such that $\pi_i > 0$, $\sum_i \pi_i a_i = 1$ and $\sum_i \pi_i \alpha_i \le w(G)$. Now if B is any maximum size clique in G then $\chi^B \in Q(G)$ and hence

$$w(G) = |B| = 1 \cdot \chi^B = \left(\sum_i \pi_i a_i^T\right) \cdot \chi^B$$
$$= \sum \pi_i (a_i \cdot \chi^B) \le \sum \pi_i \alpha_i \le w(G) .$$

So we must have equality here, in particular

$$a_i^T \chi^B = \alpha_i \quad (i = 1, \ldots, n) .$$

Since G is perfect, we know by Theorem (3.2.10) that every facet of $Q(G)$ is of the form $x_v \geq 0$ or $\chi^A \cdot x \leq 1$, where A is a stable set of nodes. Trivially, at least one of the inequalities $a_i^T x \leq \alpha_i$ is of the second kind; say $a_i^T x \leq \alpha_i$ is just the inequality $\chi^A \cdot x \leq 1$. So for every maximum clique B in G , we have

$$\chi^A \cdot \chi^B = |A \cap B| = 1 .$$

In other words, we have found a stable set A which intersects every maximum clique in G .

From here we can conclude with elementary arguments. By the choice of $A_1 = A$ above, we have $\omega(G - A_1) \leq \omega(G) - 1$. Since $G - A_1$ is again a perfect graph, we can find in it a stable set A_2 which meets all maximum cliques of $G - A_1$, i.e. we have $w(G - A_1 - A_2) \leq w(G) - 2$. Going on in a similar manner, we can partition $V(G)$ into $\omega(G)$ stable sets, i.e. we can color $V(G)$ with $\omega(G)$ colors. Thus we have proved:

(3.2.12) Theorem. *A maximum clique and an optimum coloration of a perfect graph can be found in polynomial time.* □

3.3. Minimizing a submodular function.

Let S be a finite set and $f : 2^s \to \mathbb{R}$ any function defined on the subsets of S . We say that f is *submodular,* if the following inequality holds for each pair $X, Y \subseteq S$:

$$f(X \cup Y) + f(X \cap Y) \leq f(X) + f(Y) . \tag{3.3.1}$$

Submodular setfunctions play a central role in combinatorial optimization, analogous to the role of convex functions in discrete optimization. We cannot go into the details of their theory; we refer the reader e.g. to Lovász (1983). Let us, however, give three of the most important examples of submodular setfunctions.

I. Let S be a set of vectors in a linear space $\mathbb{F}^n$, where $\mathbb{F}$ is any field. Let, for each $X \subseteq S$, $r(X)$ denote the rank of the set X , i.e. the maximum number of members of X linearly independent over $\mathbb{F}$. Then r is a submodular setfunction. Besides the submodularity (3.3.1), the rank function r satisfies the following:

$$r(X) \leq |X| \quad \text{(subcardinality);} \tag{3.3.2}$$

$$\text{if } X \subseteq Y \text{ then } r(X) \leq r(Y) \quad \text{(monotonicity).} \tag{3.3.3}$$

A submodular, subcardinal and monotone setfunction $r : 2^s \to \mathbb{Z}_+$ defines a *matroid.* Vectors in linear spaces, or equivalently, columns of a matrix, form a matroid (that is where the name comes from). But there are many other important constructions which yield matroids. Let it suffice to mention that the edges of any graph G form a matroid, if we define the rank of a set X of edges of G as the maximum number of edges in X containing no circuit. For a comprehensive treatment of matroid theory, we refer to Welsh (1976).

Matroids give rise to a rich theory of very general min–max theorems and deep polynomial–time algorithms. Let us restrict ourselves to one of the most important results. To formulate this, we need the following definition. Let S be a finite set and r , a matroid rank function on S .

An *independent set* in (S, r) is a set $X \subseteq S$ such that $r(X) = |X|$ (in the case of our original example, this would just mean linear independence). The maximum size of an independent set is $r(S)$; an independent set of this size is called a *basis.* It is easy to find a basis in a matroid: it follows easily from submodularity that if we select elements while repeatedly taking care that the elements be independent, then we always end up with a basis.

Assume now that we have two matroids (S, r_1) and (S, r_2) on the same set S . We are looking for a largest *common* independent set of these matroids. A theorem of Edmonds (1970, 1979) states that

$$(3.3.4) \qquad \max\{|A| : r_1(A) = r_2(A) = |A|\} = \min_{X \subseteq S}\{r_1(X) + r_2(S - X)\} .$$

Because we have chosen the two matroids here appropriately, a large variety of combinatorial min–max theorems follow.

II. Let V be any finite set and $S \subseteq 2^V$. Define, for $X \subseteq S$, $f(X)$ as the number of elements of V covered by X . Then $f(X)$ is a submodular setfunction.

This setfunction is featured in the famous theorem of P. Hall. We say that S has a *system of distinct representatives* if we can assign to each $A \in S$, an element $q_A \in A$ so that $q_A \neq q_B$ if $A \neq B$. Hall's Theorem tells us that S has a system of distinct representatives if and only if $f(X) \geq |X|$ for each $X \subseteq S$, i.e. if the setfunction $f(X) - |X|$ (which is also submodular) is non–negative.

III. Let G be a graph and let, for $X \subseteq V(G)$, $\delta(X)$ denote the number of edges of G connecting X to $V(G) - X$. Then δ is a submodular setfunction. This setfunction occurs in many graph theoretic results. In particular, G is k–edge–connected if and only if $\delta(X) \geq k$ for each $X \subseteq V(G)$, $X \neq \emptyset$, $V(G)$.

A version of this example has already come up before. If G is a digraph, $s, t \in V(G)$ and $c : E(G) \to \mathbb{Q}_+$ is any capacity function on the edges of G , then we can define, for each $X \subseteq V(G) - s - t$, $\delta_c(X)$ as the capacity of the $s - t$ cut determined by $X \cup \{s\}$, i.e.

$$\delta_c(X) = \sum\{c(e) : e \text{ has tail in } X \cup \{s\} \text{ and head in } V(G) - X\} .$$

Then δ_c is a submodular setfunction, and the Max–Flow–Min–Cut Theorem asserts that the maximum value of an $s-t$ flow is the minimum of $\delta_c(X)$ for $X \subseteq V(G) - s - t$.

Our main result in this section will be an algorithm to compute the minimum of a submodular setfunction. The examples above show that this is indeed a problem which comes up very often in combinatorial optimization.

We have to add a few remarks here about how f is given. Of course, f could be given as a table of all its values. In this case, it is a trivial task to find the minimum of f: scanning this table to find the minimum value does not take essentially more time than reading the input data. To get a more meaningful problem, we shall assume that f is given as an oracle that, for each input $X \subseteq S$, returns the value $f(X)$. It also has the following guarantee:

(3.3.5)

For any two sets $X, Y \subseteq S$, the outputs satisfy
$f(X) + f(Y) \geq f(X \cup Y) + f(X \cap Y)$.
For any $X \subseteq S$, the output satisfies
$\langle f(X) \rangle \leq k$.

The *input size* of this oracle is defined as k. Then we have the following theorem (Grötschel, Lovász and Schrijver (1981)):

(3.3.6) Theorem. *Given a submodular setfunction f by an oracle, a subset $X \subseteq S$ minimizing f can be found in polynomial time.*

Proof. The idea is to reduce the problem to the minimization of a convex function over the whole unit cube. f may be viewed as defined on 0-1 vectors, i.e., on the vertices of this cube. We may assume without loss of generality that $f(\emptyset) = 0$, since otherwise we can subtract $f(\emptyset)$ from all values of f. Now we define a function $\tilde{f} : \mathbb{R}_+^S \to \mathbb{R}$ as follows. Let $c \in \mathbb{R}_+^S$. Then we can write c uniquely in the form

$$c = \lambda_1 a_1 + \lambda_2 a_2 + \ldots + \lambda_n a_n , \qquad (3.3.7)$$

where $\lambda_1, \ldots, \lambda_n > 0$, $a_1, \ldots, a_n$ are 0-1 vectors and $a_1 \geq a_2 \geq \ldots \geq a_n$. In fact, this decomposition is easily constructed. We let a_1 be the incidence vector of the support of c, and λ_1, the smallest non-zero entry of c. Then $c - \lambda_1 a_1 \geq 0$ but has more zero entries as c. Going on in this fashion, we obtain the decomposition (3.3.7).

Having found this decomposition, we set

$$\tilde{f}(c) = \lambda_1 f(a_1) + \ldots + \lambda_n f(a_n) .$$

Now it is not too difficult to verify that if f is submodular then $\tilde{f}$ is convex. Furthermore, the minimum of $\tilde{f}$ over the unit cube is attained at a vertex of the cube (this is a very special property of this convex function; usually concave functions have such properties!). Hence the minimum of $\tilde{f}$ over the unit cube is equal to the minimum of f over the subsets of S.

Thus it suffices to find the minimum of f over the unit cube. But this can be accomplished in polynomial time by Theorem (2.2.15) and the methods of Section 2.3. □

The examples above show that this theorem can be applied to finding, in polynomial time, a minimum $s-t$ cut; to checking whether a setsystem has a system of distinct representatives; to finding the maximum size of common independent sets in two matroids, etc. It is quite natural that there are more efficient special–purpose algorithms for each of these tasks. But the problem of finding a combinatorial algorithm to minimize a submodular function is still open. More exactly, Cunningham (1984) proposed such an algorithm, which is, however, only pseudopolynomial: its running time is polynomial in the values of the function but not in their input size.

It is quite natural to ask why one is concerned with the minimization of a submodular setfunction and not with its maximization. Let us take the example above, and assume that all sets in S have k elements. Consider the submodular setfunction $g(X) = f(X) - (k-1)\,|X|$. Then it is easy to see that the maximum value of $g(X)$ is the maximum number of disjoint members of S.

This example has two implications.

Even for very simple realizations of the oracle describing a submodular setfunction f, the problem of maximizing f may be NP–hard. In fact, the problem of finding the maximum number of disjoint members of a system of 3–sets is NP–hard. One can prove that in the oracle model, it takes exponential time to find the maximum of a submodular setfunction.

There are special cases of the submodular function maximization problem that can be solved by quite involved methods. Most notably, case $k = 2$ of the above example is just the matching problem for graphs. This algorithm can be extended to a larger class of submodular functions, but a real understanding of this phenomenon is still missing. For some details, see Lovász (1983).

Finally, let us remark that Theorem (3.3.6) can be extended in different ways to the case in which we want to find the minimum of f only over some subfamily of 2^S. Perhaps most notable (and certainly most difficult to show) is the fact that the minimum of a submodular setfunction $f : 2^S \to \mathbb{Q}$ over the odd cardinality subsets of S can be found in polynomial time (Grötschel, Lovász and Schrijver (1985)). As a special case we find that a minimum weight T–cut can be found in polynomial time (cf. Section 3.1).

References

L. Babai [1985], *On Lovász' lattice reduction and the nearest lattice point problem*, in Lecture Notes in Comp. Sci. 182, K. Mehlhorn, ed., 2nd Annual Symp. on Theor. Asp. Comp. Sci., Springer, Berlin, pp. 13–20.

I. Bárány and Z. Füredi [1984], *private communication.*

C. Berge and V. Chvátal [1984], *Topics on Perfect Graphs*, Annals of Discrete Mathematics 21, North–Holland, Amsterdam.

E. Bishop [1967], *Foundations of Constructive Mathematics*, McGraw–Hill, New York.

R. G. Bland, D. Goldfarb and M. Todd [1981], *The ellipsoid method: a survey*, Oper. Res., 29, pp. 1039–1081.

I. Borosh and L. B. Treybig [1976], *Bounds on positive integral solutions of linear diophantine equations*, Proc. Amer. Math. Soc., 55, pp. 299–304.

J. W. S. Cassels [1965], *An Introduction to Diophantine Approximation*, Cambridge Univ. Press, Cambridge.

——— [1971], *An Introduction to the Geometry of Numbers*, Springer, Berlin.

V. Chvátal [1975], *On certain polytopes associated with graphs*, J. Combin. Theory Ser. B, 18, pp. 138–154.

W. R. Cunningham [1984], *On submodular function minimization*, Combinatorica, 5, pp. 185–192.

L. Danzer, B. Grünbaum and V. Klee [1963], *Helly's Theorem and its relatives*, in Convexity, V. Klee, ed., Amer. Math. Soc., Providence, RI, pp. 101–180.

E. W. Dijkstra [1959], *A note on two problems in connection with graphs*, Numer. Math., 1, pp. 269-271.

J. Edmonds [1967], *Optimum branchings*, J. Res. Nat. Bur. Standards B, 71, pp. 233–240.

——— [1970], *Submodular functions, matroids and certain polyhedra*, in Combinatorial Structures and their Applications, R. Guy, H. Hanani, N. Sauer and J. Schönheim, eds., Proc. Intern. Conf. Calgary, Alb., 1969. Gordon and Breach, New York, pp. 69–87.

J. Edmonds [1973], *Edge-disjoint branchings*, in Combinatorial Algorithms, R. Rustin, ed., Courant Comp. Sci. Symp., Monterey, CA, 1972. Academic Press, New York, pp. 91–96.

—— [1979], *Matroid intersection*, in Ann. of Discrete Mathematics 4, North-Holland, Amsterdam, pp. 39–49.

J. Edmonds and E. L. Johnson [1973], *Matching, Euler tours and the Chinese postman*, Math. Programming, 5, pp. 88–124.

J. Edmonds, L. Lovász and W. Pulleyblank [1982], *Brick decompositions and the matching rank of graphs*, Combinatorica, 2, pp. 247–274.

G. Elekes [1982], *A geometric inequality and the complexity of computing volume*, Disc. and Comput. Geometry (to appear).

P. van Emde Boas [1981], *Another NP-complete partition problem and the complexity of computing short vectors in a lattice*, Rep. MI/UVA 81–04, Amsterdam.

L. R. Ford and D. R. Fulkerson [1962], *Flows in Networks*, Princeton Univ. Press, Princeton, New Jersey.

A. Frank [1981], *How to make a digraph strongly connected*, Combinatorica, 1, pp. 141–153.

A. Frank and É. Tardos [1985], *A combinatorial application of the simultaneous approximation algorithm*, Combinatorica (to appear).

M. A. Frumkin [1976], *Polynomial time algorithms in the theory of linear diophantine equations*, in Fundamentals of Computation Theory, M. Karpinski, ed., Lecture Notes in Computer Science 56, Springer, Berlin, pp. 386–392.

D. R. Fulkerson [1970], *Blocking polyhedra*, in Graph Theory and its Applications, B. Harris, ed., Proc. Advanced Seminar, Madison, WI, 1969. Academic Press, New York, pp. 93–112.

—— [1971], *Blocking and anti-blocking pairs of polyhedra*, Math. Programming, 1, pp. 168–194.

—— [1974], *Packing rooted directed cuts in a weighted directed graph*, Math. Programming, 6, pp. 1–13.

J. von zur Gathen and M. Sieveking [1976], *Weitere zum Erfüllungsproblem polynomial äquivalente kombinatorische Aufgaben*, in Komplexität von Entscheidungsproblemen, E. Specker and V. Strassen, eds., Lecture Notes in Computer Science 43, Springer, Berlin, pp. 49–71.

J.-L. Goffin [1982], *Variable metric methods, part* II: *An implementable algorithm, or the Ellipsoid Method*, preprint, McGill Univ., Montreal (Working Paper no. 82–27).

M. Golumbic [1980], *Algorithmic Graph Theory and Perfect Graphs*, Academic Press, New York.

M. Grötschel, L. Lovász and A. Schrijver [1981], *The ellipsoid method and its consequences in combinatorial optimization*, Combinatorica, 1, pp. 169–197.

M. Grötschel, L. Lovász and A. Schrijver [1984a], *Polynomial algorithms for perfect graphs*, in Topics on Perfect Graphs, C. Berge and V. Chvátal, eds., Annals of Discrete Mathematics 21, North-Holland, Amsterdam, pp. 326–356.

____ [1984b], *Geometric methods in combinatorial optimization*, in Progress in Combinatorial Optimization, W. R. Pulleyblank, ed., Proc. Silver Jubilee Conference on Combinatorics, Univ. Waterloo, 1982, Vol. 1, Academic Press, New York, pp. 167–183.

____ [1984c], *Corrigendum to our paper "The Ellipsoid method and its consequences in combinatorial optimization"*, Combinatorica, 4, pp. 291–295.

____ [1984d], *Relaxations of vertex packing*, J. Combin. Theory Ser. B (to appear).

____ [1985], *Combinatorial Optimization and the Ellipsoid Method*, Springer, Berlin (to appear).

F. John [1948], *Extremum problems with inequalities in subsidiary condition*, in Studies and Essays Presented to R. Courant on his 60th Birthday, K. O. Friedrichs, O. E. Neugebauer and J. J. Stoker, eds., Jan. 8, 1948, Interscience, New York, pp. 187–204.

R. Kannan, A. K. Lenstra and L. Lovász [1984], *Polynomial factorization and nonrandomness of bits of algebraic and some transcendental numbers*, in Proc. 16th Annual ACM Symposium on Theory of Computing, pp. 191–200.

N. Karmarkar [1984], *A new polynomial-time algorithm for linear programming*, Combinatorica, 4, pp. 373–396.

R. M. Karp and C. Papadimitriou [1980], *On linear characterizations of combinatorial optimization problems*, Proc. 21st Annual Symposium on Foundations of Computer Science, IEEE, pp. 1–9.

Ker-I Ko [1983], *On the definitions of some complexity classes of real numbers*, Math. Systems Theory, 16, pp. 95–109.

L. G. Khachiyan [1979], *A polynomial algorithm in linear programming*, Dokl. Akad. Nauk SSSR, 244, pp. 1093–1096; Soviet Math. Dokl., 20, pp. 191–194.

A. Khintchine [1935], *Continued fractions*, Moscow-Leningrad.

____ [1948], *A quantitative formulation of Kronecker's theory of approximation*, Izv. Akad. Nauk SSSR, Ser. Mat. 12, pp. 113–122. (In Russian.)

J. C. Lagarias [1983], *The computational complexity of simultaneous diophantine approximation problems*, Proc. 23^{d} IEEE Symposium on Foundations of Computer Science, pp. 32–39.

J. C. Lagarias and A. M. Odlyzko [1983], *Solving low density subset sum problems*, in Proc. 24th IEEE Symposium on Foundations of Computer Science, pp. 1–10.

A. Lehman [1965], *On the length-width inequality, mimeographed notes*, Math. Programming, 16, pp. 245–259.

A. K. Lenstra, H. W. Lenstra, Jr. and L. Lovász [1982], *Factoring polynomials with rational coefficients*, Math. Ann., 261, pp. 515–534.

H. W. Lenstra, Jr. [1983], *Integer programming with a fixed number of variables*, Math. Oper. Res., 8, pp. 538–548.

H. W. Lenstra, Jr. and C. P. Schnorr [1984], *On the successive minima of a pair of polar lattices*, preprint.

L. Lovász [1972], *Normal hypergraphs and the perfect graphs conjecture*, Discrete Math., 2, pp. 253–276.

——— [1979], *On the Shannon capacity of a graph*, IEEE Trans. Inform. Theory, IT-25, pp. 1–7.

——— [1983], *Submodular functions and convexity*, in Mathematical Programming: the State of the Art, Springer, Berlin, pp. 235–257.

——— [1985], *Some algorithmic problems on lattices*, in Theory of Algorithms, L. Lovász and E. Szemerédi, eds., Proc. Conference, Pécs (Hungary) 1984, Colloq. Math. Soc. János Bolyai (to appear).

C. L. Lucchesi [1976], *A minimax equality for directed graphs*, Ph. D. thesis, Univ. Waterloo, Waterloo, Ontario.

C. L. Lucchesi and D. H. Younger [1978], *A minimax relation for directed graphs*, J. London Math. Soc., (2) 17, pp. 369–374.

M. Mignotte [1974], *An inequality about factors of polynomials*, Math. Comp., 28, pp. 1153–1157.

A. M. Odlyzko and H. te Riele [1985], *Disproof of the Mertens conjecture*, J. Reine angew. Math. 357, pp. 138–160.

M. W. Padberg and M. R. Rao [1982], *Minimum cut-sets and b-matchings*, Math. Oper. Res., 7, pp. 67–80.

M. W. Padberg and M. R. Rao [1981], *The Russian method for linear inequalities and linear optimization*, III, NYU Grad. School WP, No. 81-39.

C. P. Schnorr [1985], *A hierarchy of polynomial time basis reduction algorithms*, in Theory of Algorithms, L. Lovász and E. Szemerédi, eds., Proc. Conference Pécs (Hungary) 1984, Colloq. Math. Soc. János Bolyai (to appear).

R. Schrader [1982], *Ellipsoidal methods*, in Modern Applied Mathematics – Optimization and Operations Research, B. Korte, ed., North–Holland, Amsterdam, pp. 265–311.

P. D. Seymour [1980], *Decomposition of regular matroids*, J. Combin. Theory Ser. B, 28, pp. 305–359.

N. Z. Shor [1970], *Convergence rate of the gradient descent method with dilatation of the space*, Kibernetika, 2, pp. 80–85; Cybernetics, 6, pp. 102–108.

É. Tardos [1985], *A strongly polynomial minimum cost circulation algorithm*, Combinatorica, 5, pp. 247–255.

——— [1985], *A strongly polynomial algorithm to solve combinatorial linear programs*, Math. Oper. Res. (to appear).

A. M. Turing [1937], *On computable real numbers, with an application to the Entscheidung problems*, Proc. London Math. Soc., 42, pp. 230–265.

A. A. Voytakov and M. A. Frumkin [1976], *An algorithm for finding the general integral solution of a system of linear equations*, in Issledovaniya po Diskretnoi Optimalizatsic, A. Fridman, ed., Nauka, Moscow, pp. 128–140. (In Russian.)

D. Welsh [1976], *Matroid Theory*, Academic Press, New York.

D. B. Yudin and A. S. Nemirowskii [1976], *Informational complexity and effective methods of solution for convex extremal problems*, Ekonomika i Mat. Metody, 12, pp. 357–369; Matekon: Transl. of Russian and East European Math. Economics, 13, pp. 24–25.